Springer Monographs in Mathematics

Springer
Berlin
Heidelberg
New York
Barcelona
Budapest
Hong Kong
London
Milan
Paris
Singapore
Tokyo

Robert L. Griess, Jr.

Twelve Sporadic Groups

Springer

Robert L. Griess, Jr.
Department of Mathematics
University of Michigan
2072 East Hall
Ann Arbor, MI 48109-1109
USA
e-mail: rlg@math.lsa.umich.edu

Library of Congress Cataloging-in-Publication Data

Griess, Robert L., 1945-
Twelve sporadic groups / Robert Griess, Jr.
p. cm. -- (Springer monographs in mathematics)
Includes bibliographical references and index.

1. Finite simple groups. I. Title. II. Series.
QA177.G75 1998
512'.2--dc21 98-18097
CIP

Mathematics Subject Classification (1991):20B20, 20B25, 20D05, 20E32, 20F29

ISBN 978-3-642-08305-1

Table of Contents

Introduction

The finite simple groups are basic objects in algebra since many questions about general finite groups can be reduced to questions about the simple groups. Finite simple groups occur naturally in certain infinite families, but not so for all of them: the exceptions are called *sporadic groups*, a term used in the classic book of Burnside [Bur] to refer to the five Mathieu groups. There are twenty six sporadic groups, not definitively organized by any simple theme. The largest of these is the monster, the simple group of Fischer and Griess, and twenty of the sporadic groups are involved in the monster as subquotients. These twenty constitute the Happy Family, and they occur naturally in three generations. In this book, we treat the twelve sporadics in the first two generations. I like these twelve simple groups very much, so have chosen an exposition to appreciate their beauty, linger on details and develop unifying themes in their structure theory.

Most of our book is accessible to someone with a basic graduate course in abstract algebra and a little experience with group theory, especially with permutation groups and matrix groups. In fact, this book has been used as the basis for second-year graduate courses. We assume that the reader has had a first graduate course in algebra, including group theory and basic theory of permutation groups; the various canonical form theorems for a linear transformation over a field; the Smith canonical form for integer matrices and the interpretation by free abelian groups and subgroups; basic theory of finite fields and elementary Galois theory; basic definitions and properties of bilinear forms; and the basic theory of the alternating and symmetric groups.

Permutation representations of groups are essential throughout this book, and basics of this theory are reviewed in Chapter 1, which surveys other results from general finite group theory and a few nonstandard results.

Chapter 2 surveys results about particular finite groups, including basic group cohomology, interpretations by group extensions and calculations. In parts of this chapter, we include more material than is really needed for applications in this book. The material from homological algebra in Chapter 2 is a rapid survey, which the inexperienced reader can simply take on faith (or skip completely the first time through since most results in the book do not require this material).

The heart of the book consists of Chapters 3 through 10. Chapters 3 and 4 set up a theory of codes, Chapters 5 through 7 deal with the five Mathieu groups, Chapter 8 sets up lattice theory and Chapters 9 and 10 deal with the Leech lattice and subgroups of its automorphism group.

Chapters 3 through 9 come close to satisfying our ideal of "complete definitions and proofs". Moreover, we strive to present "elementary" proofs throughout; a few points require some homological algebra (summarized in Chapter 2) and modular function theory; in Chapter 6, we discuss conjugacy classes of maximal subgroups for the large Mathieu group M_{24} (but do not prove that every maximal subgroup is represented here). In contrast, Chapter 10 does need substantial outside material from the theories of classical groups and Chevalley groups, but such requirements are kept to a minimum by substituting elementary arguments whenever possible. Appendices give further background information, including some early results from the classification of finite simple groups applicable to the simple groups we encounter.

Chapter 11 finishes with a brief and somewhat anecdotal account of the eight remaining members of the Happy Family, plus the six Pariahs.

Most results of this book have been well-known in some form for a long time, though it is hard to find written proofs for certain points. The rectangular diagrams for the hexacode-Golay code theory, lists of subgroups for M_{24} and Co_1 are summarized in the Atlas and appear explicity in famous articles of Conway from the late 60s; my $L_2(7)$-labeling (6.29) may be new. I do make some original contributions here: the systematic use of labeling procedures and uniqueness results, the greater rigor in aspects of the theory, the full treatment of the ternary Golay code and automorphism group (Chapter 7), and the direct and elementary determination of conjugacy classes of elements of small order in certain simple groups.

Acknowledgements

I thank Dave Benson for giving me verbal sketches of the hexacode-Golay code theory in 1983; it had not been widely publicized before that time, though I heard John Conway's brief account of it in a 1981 lecture. This theory has origins in the work of Robert Curtis and was further developed during the construction of the Janko group J_4, completed in spring 1980 by Richard Parker, John Conway, David Benson, J. G. Thackray and Simon Norton [No80]. I also thank those who attended my graduate course at Yale and my several courses on this material at the University of Michigan; their comments were greatly appreciated and significantly helped to improve this text; I would like especially to mention Walter Feit, Mike Kantor, Tom Richardson and Alex Ryba. For help with certain points in the text, I thank Donald Higman, Richard Lyons, Ronald Solomon and Sia Wong. My gratitude also goes to Peter Johnson for his many critical remarks on earlier versions. Those many earlier versions were written nicely with the Chiwriter mathematics word processing package. Finally, I thank George Glauberman heartily for his extended interest in this material, his agressive and careful commentary and use of my notes in his advanced courses at the University of Chicago and lectures at Kansas State University; his help was most generous and promoted many significant improvements.

Notation for Simple Groups

My system of notation for the simple groups is the following. For Chevalley groups, I use the system in [Car]. For the alternating groups, I use Alt_n rather than A_n to avoid possible confusion with groups of Lie type A_n.

For classical groups, I use the standard symbols $GL(n,q)$, $SL(n,q)$, $O^\epsilon(n,q)$, $SO^\epsilon(n,q), \ldots$ (or the variants $GL_n(q)$, $SL_n(q)$, $O_n^\epsilon(q)$, $SO_n^\epsilon(q), \ldots$) for the general linear group, special linear group, orthogonal group, special orthogonal group, ...of dimension n over the finite field of q elements. For projective versions of these (the quotient modulo the center), I prefix with a P to get $PGL(n,q)$, etc. Now, let V be a finite dimensional vector space V with quadratic form. We define $\Omega(V)$ as subgroup of the special orthogonal group which is the kernel of the spinorial norm in characteristic not 2 [Artin] and as the kernel of the Dickson invariant [Dieu] when the field is perfect of characteristic 2 (e.g. a finite field).

The use of $O^\epsilon(n,q)$ in [Atlas] for the *nonabelian simple composition factor of the orthogonal group* is, in my opinion, a bad choice, because of possible confusion. The symbol $O^\epsilon(n,q)$ for orthogonal groups has been well established in finite group theory as and, throughout the mathematics community, $O(n,K)$ stands for an orthogonal group when K is the real or complex field. There seems little advantage to this Atlas notation. The traditional symbol $P\Omega^\epsilon(n,q)$ is perfectly reasonable for this simple composition factor, so I recommend continuing to use it.

For sporadic groups, style of notation has varied over the years; no system is fully satisfactory. At this time, I use a symbol which refers to the name(s) of the discoverer(s), possibly with a numerical subscript. An exception is the term monster, denoted $\mathbb{M}$ or F_1, and the groups F_3 and F_5 (these were easily seen within the putative simple group, the Monster, in November, 1973; coincidentally, Koichiro Harada had contemporaneously been working on a classification problem which was leading to F_5, and so could be considered its "discoverer"). The symbol $\mathbb{M}$ (my idea, proposed at the Columbus meeting in 1993) has gained acceptance as a symbol for the monster; a plain roman or italic M seems inappropriate for the permanent name of a group since it ties up use of that letter. My preferred notations for sporadic groups are all listed in (2.14) and Chapter 11.

Some other comments on Atlas material will be found in Chapter 11 and an Appendix.

Chapter 1. Background from General Group Theory

We discuss the background in group theory used in this book and present the notation and conventions in force. Unless explicitly stated otherwise, rings are assumed to be associative with unit and modules are assumed unital. There is a summery of group theoretic notation on p. ???

We begin with a review of some elementary concepts from the theory of permutation groups, without proofs, and discuss a few examples. For a basic introduction to the theory of groups acting on sets, see any basic text on group theory, e.g., [Go68], [Hu]. The book [Hu] contains a discussion of several important examples coming from classical groups. See also [Wie].

Our groups act on the right. A *permutation group* is a group together with a faithful permutation representation; sometimes, it means the image of a group under such a permutation representation.

(1.1) Definition. Let G be a group acting on a set Ω. The *degree* of the permutation representation is $|\Omega|$. Let k be a positive integer, $k \leq |\Omega|$. We say that G is *k-fold transitive* on Ω if, given any two ordered k-tuples $(x_1, \ldots, x_k)$ and $(y_1, \ldots, y_k)$ of distinct points from Ω, there is an element $g \in G$ such that $x_i^g = y_i$, for $i = 1, \ldots, k$. We say that G is *sharply k-fold transtive* if the element g postulated to exist is unique. For points $x_1, \ldots, x_m$ of Ω, $G_{x_1,\ldots,x_m} = \{g \in G | x_i^g = x_i$, for $i = 1, \ldots, m\}$ denotes the *stabilizer* in G of $x_1, \ldots, x_m$.

(1.2) Definition. Let G be a group acting transitively on a set Ω. A *system of imprimitivity* on Ω is a partition of Ω into nonempty sets, called *blocks*, such that whenever B is a block and $g \in G$, $B \cap B^g = \emptyset$ or B. This permutation representation of G is called *primitive* if the only systems of imprimitivity are the trivial ones, namely the 1-sets and Ω.

Note that we may deduce transitivity from the other properties of primitivity if Ω does not have cardinality 2.

(1.3) Lemma. (i) *A doubly transitive group is primitive.* (ii) *A nonidentity normal subgroup of a primitive group is transitive.* (iii) *If G is k-fold transitive of degree n and $x_1, \ldots, x_k$ are k distinct points,* $|G: G_{x_1,\ldots,x_k}| = n(n-1)\ldots(n-k+1)$.

(1.4) Proposition (Jordan's Theorem). *If G is a primitive permutation group of degree $n \geq 5$ and an element of G acts as a p-cycle, where p is a prime, $p \leq n-3$, then G contains the alternating group of degree n.*

(1.5) Lemma (the Frattini argument). *Let K be a normal subgroup of the group G and suppose that P is a subgroup of K whose K-conjugacy class is invariant under conjugation by G. Then $G = K \cdot N_G(P)$.*

The classic version of this result concerns the case where $P \in Syl_p(K)$. The above approximates many variations in the literature are called *Frattini arguments.*

(1.6) Exercises

(1.6.1) Show by example that the conclusion of 2.4 may not hold when we relax the hypothesis to $p \leq n - 2$.

(1.6.2) The Mathieu group M_{24} is a 5-transitive group on 24 points (accept this for the moment) Its order is $2^{10}3^3 5.7.11.23$. Show that it is not 6-transitive.

(1.6.3) The action of a 4-fold transitive group on unordered pairs of points is primitive, except for the natural representation of Σ_4. If the group is only 3-transitive, this may not be so.

(1.6.4) The action of $GL(n, q)$ on the set of nonzero vectors of $\mathbb{F}_q^n$ is primitive if and only if $n > 1$ and $q = 2$ or $n = 1$ and $q - 1$ is a prime number.

(1.6.5) Let G be the subgroup of $GL(n, q)$ stabilizing a nondegenerate quadratic, symmetric or alternating bilinear or hermitian form on $\mathbb{F}_q^n$. Use Witt's theorem to find the orbits of G on the set of nonzero vectors. Formulate and prove primitivity statements which are analogues of those in problem (1.6.4).

(1.6.6) For each $k \in \{0, 1, \ldots, n\}$, the action of $GL(n, q)$ on the set of k-dimensional subspaces of $\mathbb{F}_q^n$ is primitive; the action is doubly transitive if and only if $k = 1$ or $n - 1$.

(1.7) Definition. Let R be a ring and Ω a set on which the group G acts. The associated *permutation module* is the R-free R module $R\Omega$ with basis Ω and G action $\sum_{x \in \Omega} a_x x \mapsto \sum_{x \in \Omega} a_x x^g$, for $g \in G$. The element $\nu = \sum_{x \in \Omega} x$ is called the *norm element* of $R\Omega$. The submodule $R\Omega^o$ of $R\Omega$ defined by $\sum_{x \in \Omega} a_x = 0$ is called the *augmentation submodule* or, when R has characteristic 2, the *submodule of even elements.*

Note that the bilinear form for which Ω is an orthonormal basis is G-invariant and that the augmentation submodule is the annihilator of $R\nu$, and conversely. Note also that the augmentation submodule is spanned over R by all $x - y$, for $x, y \in \Omega$ and that we have a basis consisting of all $x - y$, where $x \in \Omega$ is fixed and $y \in \Omega \setminus \{x\}$.

(1.8) Lemma. *Let $R\Omega^o$ be the augmentation submodule and assume that R is a field of characteristic $p \geq 0$. Then:*

(i) *p divides $|\Omega|$ if and only if $\nu \in R\Omega^o$.*

(ii) *If G is the symmetric or alternating group on Ω, the only proper and nontrivial submodule of $R\Omega^o$ is $R\nu$ if $\nu \in R\Omega^o$, unless R has a primitive cube root of unity and* (a) *$G = Alt_3$ and $p \neq 3$ or* (b) *$G = Alt_4$ and $p = 2$.*

(iii) *The only nontrivial submodules of $R\Omega$ for such a G are $R\nu$ and $R\Omega^o$, except for* (a) *and* (b) *above and for* (c) *$G = A_2$ or Σ_1 (in which case $R\Omega^0 = 0$).*

(1.8.1) Exercise. (a) Show that if V is a subspace of $R\Omega$ and $dim\, V \geq 2$, then for each $x \in \Omega$, there is an element $\sum a_y y \in V$ with $a_x = 0$. (b) Find an example for each exception in (1.8).

Proof of (1.8). The first statement is obvious. Let $M \neq 0$ be a nontrivial submodule of $R\Omega^o$, $M \neq R\nu$. Then $|\Omega| \geq 3$ and so G is transitive. Let $\xi = \sum a_x x$ be a vector in M with minimal support, say S. Then $|S| \geq 2$.

Suppose $S = \Omega$. By (1.8.1.a), $dim\, M = 1$ and so G' centralizes M. For each $g \in G'$, we have $\xi = \xi^g = \sum a_x x^g$ and so $a_x = a_{x^g}$, for all $g \in G'$. Since $M \neq R\nu$, the coefficients are not constant, whence G' is not transitive on Ω. Thus, $G = Alt_3$, cyclic of order 3. Since G is transitive on Ω, $R\nu$ is the space of fixed points and so G acts nontrivially on M. Therefore, R has a primitive cube root of unity.

If S is a 2-set, we are done, due to transitivity of G on 2-sets and the fact that $R\Omega^o$ is spanned by all $x - y$, for $x, y \in \Omega$. We may assume S has more than 2 elements. If S is a proper subset of Ω whose complement has at least 2 elements, high transitivity of G produces a nonzero element of the form $\xi - \xi^g$ in M with support a 2-set (proof: take x, $y \in S$; since G is $(n-2)$-fold transitive and $n \geq 5$, there is $g \in G$ which takes x to y and is trivial on $S \backslash \{x\}$; use this g). Suppose that the complement of S is a 1-set, say $\{z\}$. If G is at least 3-transitive, then we take 3 points of S, say, w, x and y, we take an element g which satisfies $z^g = y$, $w^g = w$ and $x^g = x$. Then $\xi - \xi^g$ is nonzero with at least two zero coordinates, and we are in an earlier case. If S and G fail to satisfy the above requirements, G is only doubly transitive and S has three elements (i.e., $G = Alt_4$) or G is not even 2-transitive, i.e., $G = Alt_3$.

We dispose of the case $G = Alt_4$ as follows. Write $abc0$ for the coordinates of ξ. By commutating with the 3-cycle (123),we get a cube root ω of unity such that $b = \omega a$ and $c = \omega^2 a$ (otherwise $\xi - ba^{-1}\xi^{(123)}$ is nonzero with smaller support than ξ). Without loss, $a = 1$. By applying powers of (234) to $1\omega\bar{\omega}0$ we get vectors $10\omega\bar{\omega}$ and $1\bar{\omega}0\omega$. If $p \neq 2$ or $\omega = 1$, this is a linearly independent set, so we get $dim\, M = 3$, or $M = R\Omega^o$, as required. When $p = 2$ and $\omega \neq 1$, we have the exception as indicated in the statement of the lemma.

Suppose $G = Alt_3$. If $p = 3$, the result is clear by considering the Jordan canonical form for a generator of G and if $p \neq 3$, we get a proper submodule iff the polynomial $x^2 + x + 1$ has a root in R.

To prove the last statement, it suffices to show (except for the excluded cases) that if M is a submodule not contained in $R\nu$, it contains $R\Omega^o$. We leave the cases $|\Omega| \leq 3$ as exercises, and so we assume $|\Omega| \geq 4$. Suppose that M has an element ξ with nonzero coordinate sum. If every such ξ has equal coordinates,

$M = R\nu$. We may therefore assume that ξ has distinct coordinates at, say, x and y. Take $z \in \Omega \backslash \{x, y\}$ and $g \in G$ with $x^g = y$ and $z^g = z$. Then $\xi - \xi^g$ is in M, has coordinate sum 0, is nonzero but with a zero coordinate (hence is not a multiple of ν), whence M contains $R\Omega^0$. This suffices to complete the proof.0

(1.9) (The Schur-Zassenhaus Theorem; *see* [Go68] 6.2**).** *Let G be a finite group with normal subgroup N and quotient G/N of relatively prime orders. There exists a complement to N (a subgroup C such that $G = NC$ and $N \cap C = 1$). If N or G/N is solvable, all complements are conjugate.*

In 1963, Feit and Thompson [FT] proved that groups of odd order are solvable; so, the conjugacy of complements in (1.9) does hold.

(1.10) Corollary. *Let H and A be groups of relatively prime orders, that H or A is solvable and that A acts on H. For any prime p, A normalizes a Sylow p-subgroup of H. If P and Q are Sylow p-subgroups of H normalized by A, there is $c \in C_H(A)$ such that $P^c = Q$.*

(1.11) Definition. G be a finite group, p a prime and $P \in Syl_p(G)$. We say that a subgroup K of G is a *normal p-complement* iff K is normal, $K \cap P = 1$ and $G = KP$. Such a K exists iff $G = O_{p'}(G)P$, and, in this case, $K = O_{p'}(G)$.

(1.12) Theorem (Burnside's normal p-complement theorem; [Go68], p. 252. *Let G be a finite group, p a prime and $P \in Syl_p(G)$. Suppose that $P \leq Z(N_G(P))$. Then G has a normal p-complement.*

(1.13) Theorem (Thompson's transfer theorem). *Let G be a finite group, $T \in Syl_2(G)$ and $x \in G$ an involution. Let U be a normal subgroup of T such that T/U is cyclic. If x is in every normal subgroup N of G such that G/N is an abelian 2-group, then x is conjugate to an element of U.*

(1.14) Remarks. The theorem is proven by studying the transfer of G to T/U. The last hypothesis of the theorem is automatic if G is perfect. The hypothesis that $|x| = 2$ may be weakened to: $x^2 \in U$ and, whenever $g \in G$ and r is an even integer such that $x^{rg} \in T$, then $x^{rg} \in U$.

(1.15) Definition. The *Frattini subgroup* $\Phi(G)$ of a group G is the set of non-generators, i.e., $\{x \in G \mid$ whenever Y is a subset such that $G = \langle Y \cup \{x\}\rangle$, then $G = \langle Y \rangle\}$.

For a finite group, this is the same as the intersection of all maximal subgroups. For a finite p-group, this is the unique minimal subgroup which is normal and gives an elementary abelian quotient group; it equals the product of the commutator subgroup with the subgroup generated by all p-th powers.

(1.16) Theorem (Burnside basis theorem). *Let P be a finite p-group and α an automorphism of order prime to p. Then $\alpha = 1$ iff α induces the identity on $P/\Phi(P)$.*

(1.17) Corollary. *Let φ be the natural map $Aut\,(P) \to Aut\,(P/\Phi\,(P))$. The kernel of φ is a p-group. If $|P/\Phi\,(P)| = p^d$, the order of $Aut\,(P)$ has the form $p^c m$, where $c \geq 0$ and m divides the p'-part of $|\,GL\,(d,p)\,| = \prod_{i=0}^{d-1} (p^d - p^i)$.*

(1.18) Lemma. *Let $n > 1$ and let $m = n$ or $n+1$. In Σ_m, the centralizer of an n-cycle x is $\langle x \rangle$. The normalizer is a split extension $\langle x \rangle U$, where U is isomorphic to the group of units of the ring of integers modulo n.*

(1.19) Lemma. *$GL\,(n,F)$ is generated by the permutation matrices and the upper triangular group. Also, $GL\,(n,F)$ is generated by the monomial matrices and a single subgroup of the form $\{I + tE_{ij} | t \in F\}$, for a fixed pair of distinct indices i, j. (Such a subgroup is called a* root group*).*

Proof. The second statement and the first are equivalent since conjugates of the above root group generate the upper triangular group. We argue by induction on n. Let G^* be the subgroup so generated.

For a partition $\mathbf{p} = (n_1, n_2, \ldots)$ of n, write $GL\,(\mathbf{p}, F) = GL\,(n_1, n_2, \ldots; F)$ for the invertible n by n matrices over F which are block sums of matrices of sizes $n_1, \ldots$ down the main diagonal. Write $\Sigma\,(\mathbf{p})$ for the corresponding permutation matrices and let Σ_n denote $\Sigma\,((n))$.

Let $A \in GL\,(n,F)$. There exists a permutation matrix $P \in \Sigma_n$ such that PA has nonzero $(1,1)$-entry. Let M be a lower triangular matrix such that MPA has first column with all zeroes except for a single 1 in the $(1,1)$-position. Now let N be an upper triangular matrix such that $MPAN$ has in its first row and column with all zeroes except for a 1 in the $(1,1)$-position. Now just observe that there is a permutation matrix Q such that $Q^{-1}MQ$ is lower triangular. It follows that $G^*AG^* \cap GL\,(1, n-1; F)$ is nonempty. Since $GL\,(1, n-1; F) \leq G^*$ by induction, we are done.

(1.20) Definition. Suppose G is a group and A and B are right G-sets (i.e., there are permutation representations of G on A and B). The set $M := Maps\,(A,B)$ of functions from A to B has the following structure as a G-set: given $f \in M$, $x \in G$, $f^x \in M$ is defined by one of the formulas:

$$f^x\,(a) := f\left(a^{x^{-1}}\right)^x \text{ for } a \in A; \text{ or} \tag{1.20.1}$$

$$(a)\,f^x := \left(\left(a^{x^{-1}}\right) f\right)^x, \text{ for } a \in A, \tag{1.20.2}$$

according to whether maps are applied on the left or the right. By a change of variable, we see that these formulas are equivalent to, respectively.

$$f^x\,(u^x) = f\,(u)^x, \text{ for } u \in A; \tag{1.20.3}$$

$$(u^x)\,f^x = ((u)\,f)^x, \text{ for } u \in A. \tag{1.20.4}$$

A function in M is a G-map if iff it commutes with the action of G; these last two formulas tell us that

(1.20.5) $\quad f \in M$ is a G-map iff $f = f^x$, for all $x \in G$.

That our definition of the action of G on M is natural is corroborated by (1.20.5).

(1.21) Definition *(The associated HI-representation; ("HI" refers to "homogeneous elements with indices"))*. Let V be an RG-module for a commutative ring R and group G. Suppose that $\{V_i | i \in I\}$ is a set of R-subspaces which are permuted by the action of G and suppose that $\{p_i \mid i \in I\}$ is a set of R-maps $V \to V_i$ such that $p_i^2 = p_i$ (so that p_i is a projection) and such that the correspondence $V_i \leftrightarrow p_i$ is a G-equivariant isomorphism (see (1.20)). Let $\{X_i \mid i \in I\}$ be a family of disjoint sets, pairwise disjoint such that X_i is set-isomorphic to V_i, by $\vartheta_i : V_i \to X_i$, for all $i \in I$. Let X be the (disjoint) union of the X_i. For instance, we may take X to be the set of all pairs (h, i), where $i \in I$ and $h \in V_i$. We define an action of the group VG (semidirect product) on X as follows:

(1.21.1) for $x \in X_i$, $v \in V$, $g \in G$, $x^{vg} := \left(x^{\vartheta_i^{-1}} + v\right)^{p_i g \vartheta_j}$, if j is that index such that $V_i^g = V_j$.

Using the first description of X, we write out the argument that we have a permutation representation. Let v, $v' \in V$ and let g, $g' \in G$; suppose as above that $V_i^g = V_j$, that $x \in X_i$ and that $V_j^{g'} = V_k$. Then $x^{vgv'g'} = \left(x^{\vartheta_i^{-1}} + v\right)^{p_i g \vartheta_j v' g'} = \left(\left(x^{\vartheta_i^{-1}} + v\right)^{p_i g} + v'\right)^{p_j g' \vartheta_k} = \left(\left(x^{\vartheta_i^{-1}} + v\right)^{p_i} + v'^{g^{-1}}\right)^{g p_j g' \vartheta_k} = \left(\left(x^{\vartheta_i^{-1}} + v\right)^{p_i} + v'^{g^{-1}}\right)^{p_i g g' \vartheta_k}$; since $p_i^2 = p_i$, this equals $\left(x^{\vartheta_i^{-1}} + v + v'^{g^{-1}}\right)^{g p_j g' \vartheta_k} = x^{v v'^{g^{-1}} g g'}$, which is just what we need.

Applications of this occur in Sections 5 and 7, where G is a subgroup of the extended monomial group on V and where the V_i are summands in a decompostion $V = \oplus V_i$ and the p_i are the associated projection maps. It is an exercise with (1.21) to verify that $V_i \leftrightarrow p_i$ is indeed a G-equivariant isomorphism of G-sets.

(1.22) Definition. A *projective plane* is a set of points with a a collection of subsets, called lines, satisfying these axionms:

(1.22.1) Any two distinct points are contained in one and only one line.

(1.22.2) Any two distinct lines intersect in just one point.

(1.22.3) There are four points, no three of which are collinear (in a common line).

The containment concept may be replaced with that of an incidence relation.

(1.23) Theorem. *Suppose that the projective plane is finite. There is an integer $n \geq 2$ such that every point is on $n + 1$ lines and every line contains $n + 1$ points. The number of points and the number of lines is $n^2 + n + 1$.*

The integer n is called the *order* of the projective plane. It seems to be an open question whether n must be a prime power.

Chapter 2. Assumed Results about Particular Groups

We hope that the reader has some familiarity with elementary homological algebra. We begin with a survey of elementary useful results from group extension theory. Later, we focus on particular extensions which come up in our theory of sporadic groups. For general reference, we suggest [Ben], [Gru], [HiSt], [MacL], [Rot].

(2.1) Definition. A *group extension* is a short exact sequence of groups $1 \to K \xrightarrow{\iota} E \xrightarrow{\pi} G \to 1$, is called *an extension of* G *by* K or an extension of K by G (practice varies: if G is nonsolvable and K abelian, usually we say an extension of G by K; if K is nonsolvable and G is solvable, we say an extension of K by G). The extension is *split* if there is a group homomorphism $\sigma: G \to E$ such that $\sigma\pi = 1$; otherwise, *nonsplit.*

Two extensions of G by K are called *equivalent* if there are vertical homomorphisms as indicated such that the diagram

$$\begin{array}{ccccccccc} 1 & \longrightarrow & K & \xrightarrow{\iota} & E & \xrightarrow{\pi} & G & \longrightarrow & 1 \\ & & \downarrow & & \downarrow & & \downarrow & & \\ 1 & \longrightarrow & K & \xrightarrow{\iota_1} & E_1 & \xrightarrow{\pi_1} & G & \longrightarrow & 1 \end{array}$$

commutes (the outer vertical arrows are the identity maps).

We call K the *extension kernel* and G the *quotient of the extension.* In any case, E operates on K by conjugation and so we have a homomorphism $E \to Aut\,(K)$. In case K is abelian, it is in the kernel of this map and we actually have a homomorphism from $G \cong E/K$ to $Aut\,(K)$, whence K acquires the structure of a G-module.

We use the shorthand notation $K.G$ for such a group E; the extension type will be understood from context. In case the extension is split, we write $K{:}G$ and if nonsplit, we write $K{}^{\cdot}G$.

(2.2) Examples. (i) If E is an elementary abelian p-group, the extension splits. If $K \cong \mathbb{Z}_4$ and $G \cong \mathbb{Z}_2$, E is one of four groups (two abelian ones and the dihedral and quaternion groups of order 8); we have splitting in case $E \cong \mathbb{Z}_4 \times \mathbb{Z}_2$ or Dih_8 only.

(ii) If $K \cong \mathbb{Z}_2$ and $G \cong \mathbb{Z}_2 \times \mathbb{Z}_2$, there are eight equivalence classes of such E; their isomorphism types are $\mathbb{Z}_2^3$, $\mathbb{Z}_2 \times \mathbb{Z}_4$ (three times), $Quat_8$ and Dih_8 (three times).

(iii) Let K and E be dihedral of orders 2^n and 2^{n+1}, respectively, for $n \geq 2$. The extension splits. Let $\Sigma = \{\sigma : G \to E \mid \sigma \text{ is a splitting homomorphims}\}$. The set Σ consists of 2^{n-1} maps, which form one equivalence classes in the sense that $\sigma \sim \sigma'$ iff there is an inner automorphism α of K such that $\sigma = \sigma' \left(\alpha|_{Im(\sigma')}\right)$. In case K is cyclic of order 2^n and E is dihedral, Σ is the union of two equivalence classes.

In the case of abelian extension kernels, the theory of cohomology of groups is relevant. We give a brief sketch of the theory here and refer to (2.6) for interpretations relevant to extension theory.

(2.3) Notation. R is a commutative ring, G is a group, $\mathcal{M}$ is the category of right RG-modules and A is the category of R-modules. The morphism sets in both categories are naturally abelian groups and we consider *additive* functors, i.e., those which preserve sums of morphisms (hence which carry direct sums to direct sums).

(2.4) Definition *(Cohomology of groups)*. Let H^n, $n = 0, 1, \ldots$ be a sequence of functors $\mathcal{M} \to A$ which satisfy these axioms:

(2.4.1) $H^0 M = M^G$, the fixed points, i.e., $\{x \in M \mid xg = x, \text{ for all } g \in G\}$;

(2.4.2) *(eraseability)* given M, there is a monomorphism $0 \to M \to \mathrm{I}$ such that $H^n M \to H^n I$ is the 0-map.

(2.4.3) *(the long exact sequence)* given a short exact sequence $0 \to M' \to M \to M'' \to 0$, there are natural transformations $\delta = \delta^n : H^n M'' \to H^{n+1} M'$ such that, the following infinite sequence, with maps coming from the functorality of the H^n and δ, is exact:

$$0 \to H^0 M' \to H^0 M \to H^0 M'' \xrightarrow{\delta} \ldots \to H^n M' \to H^n M \to$$

$$H^n M'' \xrightarrow{\delta} H^{n+1} M' \to H^{n+1} M \to H^{n+1} M'' \xrightarrow{\delta} \ldots .$$

(Strictly speaking, we must interpret the δ^n as natural transformations of functors on the category of short exact sequences in $\mathcal{M}$ rather than on $\mathcal{M}$ itself).

(2.4.4) *(naturality)* suppose that we are given two short exact sequences $0 \to M' \to M \to M'' \to 0$ and $0 \to Q' \to Q \to Q'' \to 0$ together with a morphism between them:

$$\begin{array}{ccccccccc} 0 & \longrightarrow & M' & \longrightarrow & M & \longrightarrow & M'' & \longrightarrow & 0 \\ \downarrow & & \downarrow & & \downarrow & & \downarrow & & \downarrow \\ 0 & \longrightarrow & Q' & \longrightarrow & Q & \longrightarrow & Q'' & \longrightarrow & 0. \end{array}$$

Then, the following diagram, involving the long exact sequences and functorially devived vertical arrows, commutes:

$$\begin{array}{ccccccccccccc} \ldots & \longrightarrow & H^{n-1}M'' & \longrightarrow & H^n M' & \longrightarrow & H^n M & \longrightarrow & H^n M'' & \longrightarrow & H^{n+1} M'' & \longrightarrow & \ldots \\ & & \downarrow & & \downarrow & & \downarrow & & \downarrow & & \downarrow & & \\ \ldots & \longrightarrow & H^{n-1}Q'' & \longrightarrow & H^n Q' & \longrightarrow & H^n Q & \longrightarrow & H^n Q'' & \longrightarrow & H^{n+1} Q'' & \longrightarrow & \ldots . \end{array}$$

A sequence of functors satisfying the above axioms is unique up to natural equivalence. The notation $H^n(G, M)$ is usual for $H^n M$, and it is called *the n-th cohomology group of G with coefficients in M*. A more compact description is that these functors are the right derived functors of the fixed point functor on the category of G-modules.

(2.5) Elementary Consequences. We mention a few matters relevant to arguments in this book. (i) If $r \in R$, $rM = 0$, then $rH^n M = 0$ for all n. (ii) If $n \geq 1$ and G is finite, $|G|$ annihilates $H^n M$. (iii) If $n \geq 1$, $H^n(G, M) = 0$ if G and M are finite and $\left(|G|, |M|\right) = 1$.

(2.6) Interpretations at particular degrees. Let G be a group and A a G-module. For definitions of n-cocycles $Z^n(G, M)$ and n-coboundaries $B^n(G, M)$, see one of the suggested texts.

(2.6.0) $H^0(G, A) = A^G$.

(2.6.1) $H^1(G, A)$ corresponds to the set of conjugacy classes of complements in a split extension $A{:}G$; the correspondence depends on the choice of a fixed complement (and so the 0-element of $H^1(G, A)$ does not single out a particular conjugacy class).

We have $H^1(G, A) = Z^1(G, A) / B^1(G, A)$; given any extension $1 \to A \to E \to G \to 1$, $Z^1(G, A)$ corresponds to $\{\alpha \in Aut(E) \mid \alpha \text{ is trivial on } A \text{ and on } E/A\}$ and $B^1(G, A) = \left\{x \to a^{-1}xa \mid a \in A\right\}$.

(2.6.2) $H^2(G, A)$ classifies equivalence classes of extensions, with 0 corresponding to the split extension.

(2.6.3) $H^3(G, A)$ measures obtructions to forming certain kinds of group extensions. Suppose we are given G, a group H and a homomorphism $\pi: G \to Out(H)$. There is an associated element ξ in $H^3(G, Z(H))$ (there is a well-defined action of G on $Z(H)$ gotten by taking any set map $G \to Aut(H)$ which when composed with the natural $Aut(H) \to Out(H)$ gives π). A theorem of MacLane and Eilenberg says that the following two statements are equivalent: (1) there is a group E such that H is a normal subgroup of E, giving quotient G and such that the associated homomorphism of G to $Out(H)$ gotten from E is π; (2) $\xi = 0$. In this sense, ξ is an obstruction.

(2.6.4) For an interpretation of higher degree cohomology groups, see [Holt].

(2.7) More Examples

(2.7.1) If A is an abelian group, G acts on A, $z \in Z(G)$ satisfies: $z - 1 \in Aut(G)$, then $H^n(G, A) = 0$, for all $n \geq 0$ (this is an observation of Jack McLaughlin). This generalizes: for instance, if A and B are modules for RG and e and f are orthogonal idempotents which act as 1 on A and B, respectively, then $Ext^n(A, B) = 0$, for all $n \geq 0$.

A corollary is that cohomology of many classical groups on their standard modules is 0, e.g., $GL(n, F)$ on F^n, when F has more than 2 elements.

(2.7.2) $dim_{\mathbb{F}_2} H^1 \left(GL(n, \mathbb{F}_2), \mathbb{F}_2^n\right) = 0$ for $n \neq 3$; 1 for $n = 3$.

(2.7.3) $dim_{\mathbb{F}_2} H^2 \left(GL(n, \mathbb{F}_2), \mathbb{F}_2^n\right) = 0$ for $n \neq 3, 4, 5$; 1 for $n = 3, 4, 5$ [Bl] [De].

(2.7.4) Occasionally, one knows that $H^3(G, Z(H)) = 0$ for trivial reasons, e.g. $\left(|G|, |Z(H)|\right) = 1$, so obstructions automatically vanish. On the other hand, here is an example of a nontrivial obstruction. Let $H = SL(2,9)$, $G = Aut(H) \cong Alt_6 \cdot 2^2$. There is no group E which satisfies $1 \to H \to E \to \mathbb{Z}_2 \times \mathbb{Z}_2 \to 1$, $E/Z(H) \cong G$ via the action of E on H by conjugation, so the associated obstruction in nonzero.

(2.8) Exercises

Prove that $H^1 \left(GL(3,2), \mathbb{F}_2^3\right) = 0$.

One proof involves studing projective resolutions of the $\mathbb{F}_2 GL(3,2)$-module; for instance, see the appendix in [Ben].

Here is a method using (2.6.1). Let $E = V{:}G$. a split extension of $V \cong \mathbb{F}_2^3$ by $G \cong GL(3,2)$. Suppose that H is a complement not conjugate to G. A Sylow 7-normalizer of E, G or H is a Frobenius group of shape 7:3 and so by the conjugacy part of Sylow's theorem, we may assume that $G \cap H$ contains a Sylow 7-normalizer, say N. Since G and H are simple groups and N has index 8, it is maximal in both and so $G \cap H = N$. Let $T \in Syl_3(N)$. Then, $N_G(T) \cong N_H(Y) \cong \Sigma_3$ and $N_E(T) \cong 2 \times \Sigma_3$. In $N_E(T)$, the center is $C_V(T)$ and it is complemented by the two different subgroups $N_G(Y)$ and $N_H(T)$. Furthermore, these are the only two complements. It is therefore clear that at most two subgroups isomorphic to $GL(3,2)$ may contain N. This bounds $|H^1 \left(GL(3,2), \mathbb{F}_2^3\right)|$ above by 2.

On the other hand, let N and T be as above and let r and s be involutions which invert T and satisfy $Vr = Vs$. Then $\langle N, r\rangle$ and $\langle N, s\rangle$ are subgroups isomorphic to $GL(3,2)$ (this may be proved by a verification of a presentation for $GL(3,2) \cong PSL(2,7)$; see (2.27)).

An important special case is that of central extensions.

(2.9) Definition. An extension is *central* if the extension kernel is in the center. A central extension is a *covering* or an *essential central extension* if the kernel is in the derived group.

(2.10) Definition. Given a group G, a central extension K is a *covering group* if the extension is maximal essential, i.e., if $1 \to A \to L \to G \to 1$ is an essential central extension and there is a group homomorphism $\varphi: L \to K$ such that

$$\begin{array}{ccccccccc} 1 & \longrightarrow & A & \longrightarrow & L & \longrightarrow & G & \longrightarrow & 1 \\ & & \downarrow & & \downarrow \varphi & & \downarrow id & & \\ 1 & \longrightarrow & B & \longrightarrow & K & \longrightarrow & G & \longrightarrow & 1 \end{array}$$

commutes, then φ is an isomorphism.

(2.11) Definition. Let G be a finite group. A *representation group (Darstellungsgruppe)* is a group K which is a central extension of G such that, given any algebraically closed field F, integer $n \geq 1$, and a homomorphism $G \to PGL(n, F)$, there is a unique homomorphism $K \to GL(n, F)$ such that

$$\begin{array}{ccc} K & \longrightarrow & GL(n,F) \\ \downarrow & & \downarrow \\ G & \longrightarrow & PGL(n,F) \end{array}$$

commutes.

(2.12) Remarks. (2.12.1). In the case of G finite, a covering group exists and any such is a representation group. In case G is perfect $\left(G = G'\right)$, a covering group is unique up to isomorphism.

(2.12.2) If $1 \to A \to K \to G \to 1$ is a covering sequence for G, A is called the *Schur multiplier (Schursche Multiplikator)*. It is isomorphic to $H_2(G, \mathbb{Z})$, and so is functorial in G. In case G is finite, this is also isomorphic (but not naturally) to $H^2(G, T)$, where T is any divisible group with trivial G-action whose torsion subgroup is $\mathbb{Q}/\mathbb{Z}$, e.g., $\mathbb{Q}/\mathbb{Z}$ or $\mathbb{C}^\times$.

(2.12.3) In general, the isomorphism type of K is not uniquely determined by G. There are at most $|\,Hom\left(G/G', A\right)|$ such isomorphism types and so there is a unique covering group if G is perfect. If $G = \Sigma_n$, there are two isomorphism types of covering groups, except for $n = 6$ and $n \leq 3$, for which there is one.

(2.12.4) Let G be a group and let $1 \to R \to F \to G \to 1$ be a free presentation. By Hopf's theorem, $H_2(G, \mathbb{Z}) \cong R \cap F'/[R, F]$. If $S/[R, F]$ complements $R \cap F'/[R, F]$ in $R/[R, F]$, F/S is a covering group of $G \cong F/R$ and conversely, any covering group is realizable this way. Alperin observed that, when G is perfect, any covering group is isomorphic to $F'/[R, F]$. An easy corollary of this is the lifting of $Aut(G)$ to $Aut(K)$ when G is perfect; see [Gr73b] for details.

(2.12.5) Given G, if $\alpha: H \to G$ is an essential central extension (2.9), there is a covering group $\gamma: K \to G$ and an epimorphism $\beta: K \to H$ so that $\gamma = \beta\alpha$.

(2.13) Theorem. *Let G be a finite group and P a Sylow p-subgroup, for a prime number p. Suppose that M is a finite G-module and is a p-group. Then, for each $n \geq 1$, $H^n(G, M)$ injects into $H^n(P, M)$ and the image is the set of stable elements (see* [CE], *p.* 259*).*

(2.14) Table 2.1 (Schur multipliers of the known finite simple groups). The computations were begun by Schur, who treated the alternating groups and the groups $PSL(2, q)$. The families of groups of Lie type were done mostly by Steinberg and Griess. In each infinite family, associated to a field in characteristic p, the p'-part of the Schur multiplier is the center of a naturally occurring group from the general theory of Chevalley groups and variations; the p-part of the Schur multiplier is trivial except for finitely many groups in the family. For a summary, see [Gr85], [Gr87b]; see also [Schur04, 06, 07], [St63, 67, 81] and [Gr12a, 12b, 13b, 74, 80b].

Schur multipliers of the known finite simple groups.

Alternating groups

Alt(n)	*Multiplier*
$n \leq 3$	1
$n = 4, 5$ and $n \geq 8$	Z_2
$n = 6, 7$	Z_6

Finite Groups of Lie Type

The multiplier of $G(q)$, $q = p^n$, is $R \times P$ where R is a p'-group and P is a p-group			
Group of Lie type	R	$R \times P$	(when P is $\neq 1$)
$A_l(q)$	$\mathbb{Z}_{(l+1,q-1)}$	$\mathbb{Z}_2$	$(l, q) = (1, 4)$
		$\mathbb{Z}_2 \times \mathbb{Z}_3$	$(1, 9)$
		$\mathbb{Z}_2$	$(2, 2)$
		$\mathbb{Z}_3 \times \mathbb{Z}_4 \times \mathbb{Z}_4$	$(2, 4)$
		$\mathbb{Z}_2$	$(3, 2)$
$B_l(q)$, $l \geq 2$	$\mathbb{Z}_{(2,q-1)}$	$\mathbb{Z}_2$	$(2, 2)$
		$\mathbb{Z}_2$	$(3, 2)$
		$\mathbb{Z}_3 \times \mathbb{Z}_2$	$(3, 3)$
$C_l(q)$, $l \geq 2$	$\mathbb{Z}_{(2,q-1)}$	$\mathbb{Z}_2$	$(2, 2)$
		$\mathbb{Z}_2$	$(3, 2)$
$D_l(q)$, $l \geq 4$	$\mathbb{Z}_{(4,q-1)}$, l odd	$\mathbb{Z}_2 \times \mathbb{Z}_2$	$(4, 2)$
	$\mathbb{Z}_{(2,q-1)} \times \mathbb{Z}_{(2,q-1)}$, l even		
$E_6(q)$	$\mathbb{Z}_{(3,q-1)}$		
$E_7(q)$	$\mathbb{Z}_{(2,q-1)}$		
$E_8(q)$	1		
$F_4(q)$	1	$\mathbb{Z}_2$	$(4, 2)$
$G_2(q)$	1	$\mathbb{Z}_3$	$(2, 3)$
		$\mathbb{Z}_2$	$(2, 4)$
${}^2A_l(q)$, $l \geq 2$	$\mathbb{Z}_{(l+1,q+1)}$	$\mathbb{Z}_2$	$(3, 2)$
		$\mathbb{Z}_4 \times \mathbb{Z}_3 \times \mathbb{Z}_3$	$(3, 3)$
		$\mathbb{Z}_3 \times \mathbb{Z}_2 \times \mathbb{Z}_2$	$(5, 2)$
${}^2B_2(q) \cong \mathrm{Sz}(q)$	1	$\mathbb{Z}_2 \times \mathbb{Z}_2$	$(2, 8)$
${}^2D_l(q)$, $l \geq 4$	$\mathbb{Z}_{(4,q+1)}$		
${}^2E_6(q)$	$\mathbb{Z}_{(3,q+1)}$	$\mathbb{Z}_3 \times \mathbb{Z}_2 \times \mathbb{Z}_2$	$(6, 2)$
${}^2F_4(q)$	1		
${}^2F_4(2)'$	1		
${}^2G_2(q)$	1		
${}^3D_4(q)$	1		

Sporadic groups

Symbol for Sporadic Group (Discoverer)	Order	Multiplier
M_{11} (Mathieu's groups)	$2^4 3^2 5.11$	1
M_{12}	$2^4 3^3 5.11$	$\mathbb{Z}_2$
M_{22}	$2^7 3^2 5.7.11$	$\mathbb{Z}_{12}$
M_{23}	$2^7 3^2 5.7.11.23$	1
M_{24}	$2^{10} 3^3 5.7.11.23$	1
J_1 (Janko's groups)	$2^3 5.7.11.19$	1
$HJ = J_2$, (M. Hall and Z. Janko)	$2^7 3^3 5^2 7$	$\mathbb{Z}_2$
J_3	$2^7.3^5 5.17.19$	$\mathbb{Z}_3$
J_4 (Janko)	$2^{21} 3^3 5.7.11^3.23.29.31.37.43$	1
Held (Held)	$2^{10} 3^2 5^2.7^3.17$	1
HiS (Higman-Sims)	$2^9 3^2 5^3 7.11$	$\mathbb{Z}_2$
McL (McLaughlin)	$2^7 3^6 5^3 7.11$	$\mathbb{Z}_3$
Suz (Suzuki)	$2^{13} 3^7 5^2 7.11.13$	$\mathbb{Z}_6$
$\cdot 1$ (Conway's groups)	$2^{21} 3^9 5^4 7^2 11.13.23$	$\mathbb{Z}_2$
$\cdot 2$	$2^{18} 3^6 5^3 7.11.23$	1
$\cdot 3$	$2^{10} 3^7 5^3 7.11.23$	1
F_{22} (Fischer's 3-transposition groups)	$2^{17} 3^9 5^2 7.11.13$	$\mathbb{Z}_6$
F_{23}	$2^{18} 3^{13} 5^2 7.11.13.17.23$	1
F'_{24}	$2^{21} 3^{16} 5^2 7^3 11.13.17.23.29$	$\mathbb{Z}_3$
LyS (Lyons)	$2^8 3^7 5^6 7.11.31.37.67$	1
Ru (Rudvalis)	$2^{14} 3^3 5^3 7.13.29$	$\mathbb{Z}_2$
O'S (O'Nan)	$2^9 3^4 5.7^3.11.19.31$	$\mathbb{Z}_3$
F_2 (Fischer's $\{3,4\}$-transposition group)	$2^{41} 3^{13} 5^6 7^2 11.13.17.19.23.47$	$\mathbb{Z}_2$
F_1 (Fischer-Griess)	$2^{46} 3^{20} 5^9 7^6 11^2 13^3 17.19.23.29.31.41.47.59.71$	1
F_3 (Thompson)	$2^{15} 3^{10} 5^3 7^2 13.19.31$	1
F_5 (Harada)	$2^{14} 3^6 5^6 7.11.19$	1

(2.15) Examples. We discuss central extensions of a few groups which are relevant to later sections of this book. In what follows, G shall be a finite simple group and A its Schur multiplier.

(2.15.1) Let $G = Alt_6$; then $A \cong \mathbb{Z}_6$. The group $3 \cdot G$ is embedded in $GL(3, \mathbb{C})$ and $K = 6 \cdot G$ is embedded in $GL(6, \mathbb{C})$.

(2.15.2) Let $G = PSL(3,4)$; then $A \cong \mathbb{Z}_{12} \times \mathbb{Z}_4$ ("12×4" for short). The perfect groups $6 \cdot G$, $4 \cdot G$ (one of the two isomorphism types) and $2 \cdot G$ embed in $GL(6, \mathbb{C})$, $GL(8, \mathbb{C})$ and $GL(10, \mathbb{C})$ respectively. The containment $Alt_6 \leq G$ (see (4.5)), when lifted to the coveing group of G, gives the extension $48 \cdot Alt_6$, whose derived group is the covering group of Alt_6.

(2.15.3) The groups Alt_6 and $PSL(2,9)$ are isomorphic and have Sylow groups isomorphic to Dih_8. The outer automorphism group is $\mathbb{Z}_2 \times \mathbb{Z}_2$. Let G be the group between $Inn(PSL(2,9))$ and $Aut(PSL(2,9))$ corresponding to adjoining the product of the field automorphism and a diagonal matrix of nonsquare determinant. Then G has semidihedral Sylow 2-groups (G is isomorphic to the Mathieu group M_{10}; see Section 6). The 2-part of the Schur multiplier of G is therefore trivial (see [Hu], 25.1 and 25.2(b), applied to the presentation $\langle x, y \mid x^y = x^3, y^2 = 1\rangle$ for a semidihedral group of order 16). Any nonsplit central extension of Alt_6 by $\mathbb{Z}_2$ is isomorphic to $SL(2,9)$ (2.14), so it follows that there does not exist a group H with the property that $H' \cong 2 \cdot Alt_6 \cong SL(2,9)$ and $H/Z(H') \cong G$.

(2.15.4) We show that there is a unique group G of the form $3 \cdot \Sigma_6$, i.e., $G' = G'', G' \cong 3 \cdot Alt_6$ and $G/O_3(G) \cong \Sigma_6$. Let H be the unique group of the form $3 \cdot Alt_6$. We argue that (a) there is an outer automorphism α of H which induces on $H/Z(H)$ the automorphism coming from conjugation by a transposition; (b) if G is as above, then $G \cong H{:}\langle\alpha\rangle$. To prove (a), note that Alt_6 is perfect, so any automorphism lifts uniquely to an automorphism of a covering group, say K (2.12.4). Since $Z(K) \cong \mathbb{Z}_6$, every subgroup is characteristic, whence we get an automorphism α of $H \cong K/O_2(K)$. If $\beta \in G$ corresponds to $\alpha, \beta^2 \in O_3(G)$, which is inverted by β; thus, $\beta^2 = 1$ and so G splits over G'. Thus, $G \cong H$.

(2.16) Table. Some degree 1 cohomology groups

Table of first cohomology groups

Group	Module	Dimension H^1	Comments/ Reference
$SL(n,q)$	$\mathbb{F}_q^n$	$\begin{cases} 0 & \text{if } (n,q) \neq (3,2) \\ 1 & \text{else} \end{cases}$	[Hig62]
$Sp(2n,q)$, q even	$\mathbb{F}_q^{2n}$	$\begin{cases} 0 & \text{if } n = 2 \\ 1 & \text{else} \end{cases}$	[Po]
$\Omega^\epsilon(2n,q)$, q even	$\mathbb{F}_q^{2n}$	$\begin{cases} 1 & \text{if } n, q) = (3,2) \\ 0 & \text{else} \end{cases}$	[Po]
M_{24}	Golay code mod universe	0	[Gr74]
M_{24}	even part of Golay cocode	1	[Gr74]

(2.17) Table. Some degree 2 cohomology groups

Table of second cohomology groups

Group	Module	Dimension H^2	Comments/ Reference
$SL(n,q)$	$\mathbb{F}_q^n$	$\begin{cases} 0 & \text{if } (n,q) \neq (3,2),(4,2) \\ & (5,2),(3,3^m),\ m \geq 2; \\ 1 & \text{else} \end{cases}$	[Bl] [De] [McL75] ...
$Sp(2n,q)$, q even	$\mathbb{F}_q^{2n}$	$\begin{cases} 0 & \text{if } n = 2 \\ 1 & \text{else} \end{cases}$	[Gr73a] [McL75]
$\Omega^\epsilon(2n,q)$, q even	$\mathbb{F}_q^{2n}$	$\begin{cases} 1 & \text{if } n \geq 4 \text{ or} \\ & n = 3 \text{ and } \epsilon = - \\ 0 & \text{else} \end{cases}$	[Gr73a] [McL95]
M_{24}	Golay code mod universe	0?	
M_{24}	even part of Golay cocode	≥ 1	[Fi71c]

The following two results are useful.

(2.18) Theorem. *Let M be a G-module and let $\mathcal{L}$ be a family of subgroups of G with the properties:*

(2.18.1) *If $L \in \mathcal{L}$, $H^i(L,M) = 0$, for $i = 0, 1$;*

(2.18.2) *Given L_1 and L_2 in $\mathcal{L}$, there is $L \in \mathcal{L}$ such that $L \leq L_1 \cap L_2$.*

(2.18.3) *The subgroups in $\mathcal{L}$ generate G.*

Then $H^i(G,M) = 0$, for $i = 0, 1$.

Proof [AlpGor]. A statement as above with k ranging over $0, 1, 2$ would be false. □

(2.19) Theorem. *If G acts on the finite dimensional vector space or finite abelian group V and G has a subnormal nilpotent subgroup N with $H^0(N,V) = 0$, then $H^k(G,V) = 0$, for all $k \geq 0$.*

Proof. [Cur] [Rob]. □

(2.20) Definition. Let p be a prime. An *extraspecial p-group* is a finite p-group P such that $P' = Z(P)$ has order p.

It follows that $P/Z(P)$ is elementary abelian and so that $\Phi(P) = P' = Z(P)$ has order p. This condition is often part of the definition of extraspecial group, but need not be.

For basic properties, including representation theory, see [Go68] [Hu]. An extraspecial p-group has order p^{1+2n} for some $n \geq 1$. For a fixed n, there are two isomorphism types, denoted p_ε^{1+2n}, for $\varepsilon = \pm$. The p-th power map on an extraspecial group P induces a well-defined map q from the vector space $P/P' \to Z(P)$. For p odd, q is a homomorphism and we let $\varepsilon = +$ iff q is the trivial map (i.e., P has exponent p). For $p = 2$, q is a quadratic form (3.2).

(2.21) Definition. (i) A *holomorph* of a group G is a group E containing G as a normal subgroup such that $C_E(G) = Z(G)$ and the action of E by conjugation on G induces $Aut(G)$. In case the conjugation action induces a proper subgroup of $Aut(G)$, we call E a *partial holomorph.*

(Our use of the term holomorph replaces an older and rare usage which means the semidirect product of a group with its automorphism group).

(ii) Now let P be an extraspecial p-group of order p^{1+2n}. It is well-known that irreducible faithful representations exist and have degree p^n; furthermore, there are $p-1$ such and any two are algebraically conjugate. A holomorph is *standard* if there is a linear representation of G of degree p^n which extends an irreducible representation of P of degree p^n; otherwise the holomorph is *twisted.*

(iii) If G is a standard holomorph with associated irreducible character χ and $g \in G$, then $\chi(g)$ is 0 or a root of unity times a power of p [Gr86a]. More precisely, let z generate $Z(P)$. Then $\chi(g) = 0$ iff g is conjugate to gz in P. Otherwise, there is a p-th root of unity ε and an integer d so that $\chi(g) = \varepsilon p_1^{d/2}$; the integer d is the dimension of the fixed point subspace for the action of g on P/P' and $p_1 = \pm p$. Note that g is not conjugate to gz if g is a p'-element.

(2.22) Theorem. *Holomorphs of P exist. In the notation of* (2.21), *the irreducible representations of E which restrict to a multiple of a fixed faithful irreducible representation of P are expressible as a tensor product of two irreducible projective representations; one is a fixed representation of degree* p^n *and the second ranges over a family of projective representations of* G/P, *considered as representations of* G.

Proof. See [Gr73a] [Gr86a]. □

(2.23) Notation. K is a perfect field of characteristic 2. D and Q are 2-dimensional vector spaces with bases x, y and K-valued quadratic form q such that

$$D: \quad qx = 0,\ qy = 0,\ q(x+y) = 1$$

$$Q: \quad qx = 1,\ qy = 1,\ q(x+y) = 1$$

Also, let I be the 1-dimensional vector space with basis x and quadradic form defined by $qx = 1$. Write $A \perp B$ for the orthogonal direct sum of spaces A and B.

(2.24) Proposition. *Let K be a perfect field of characteristic* 2 *and let V be an n-dimensional vector space with quadratic form q whose associated bilinear form f has radical* 0.

(a) *if* $\dim V = 2$, *V is isometric to D or Q.*

(b) *V is isometric to an orthogonal direct sum of D's and Q's.*

(c) *If* $V \cong \perp^r D \perp^s Q \cong \perp^{r'} D \perp^{s'} Q, r \equiv r' \pmod 2$ *and* $s \equiv s' \pmod 2$; *the converse holds, also.*

(d) *If* $rad(f)$ *has dimension* 1, *but* $q|_{rad(f)} \neq 0$, *then, for any* r, s *such that* $2(r+s) = n-1$, $V \cong \perp^r D \perp^s Q \perp I$.

Proof. (a) Suppose there is $x \neq 0$ in V such that $qx = 0$. Let y satisfy $f(x,y) = 1$, $q(y) = c$. If $c = 0$, we are done, so assume that $c \neq 0$. We have $q(ay + x) = a^2c + a = a(ac+1)$, so taking $a = c^{-1}$, we get that $V \cong D$ by taking the basis x and $ay + x$.

We now suppose that no such x exists. Take any basis $\{x, y\}$, $qx = a$, $qy = b$, $f(x,y) = c \neq 0$. Since K is perfect, we may assume $a = b = 1$. Taking $y = dx + ey$, we get $q(dx + ey) = d^2 + de + e^2$.

(b) Since the associated bilinear form is nondegenerate, there are $x, y \in V$ with $qx \neq 0$ and $f(x,y) \neq 0$. Then $S := span\{x, y\}$ is isometric to D or Q by (a). Since S is a nonsingular subpspace, V is the orthogonal direct sum of S and $S^\perp$ and we finish by induction.

(c) It suffices to prove: (c.1) $Q \perp Q \cong D \perp D$; and (c.2) $\perp^r D$ has a totally singular subspace of dimension r while $\perp^{r-1} D \perp Q$ does not, for $r \geq 1$.

(c.1): Let x_i, y_i be a basis for $Q \perp Q$, $i = 1,2$, with $f(x_i, y_j) = \delta_{ij}$ and $q(x_i) = q(y_i) = 1$. The new basis $x_1 + x_2$, y_1, y_2 suggests the isometry.

(c.2): Let x_i, y_i be a basis with $f(x_i, y_j) = \delta_{ij}$ and $qx_i = qy_j = 0$. Then span $\{x_1, \ldots, x_r\}$ is a totally singular subspace.

Now let V have basis as above, except with $qx_r = qy_r = 1$ and suppose that S is a totally singular subspace. Then, $S \cap span\{x_r, y_r\} = 0$ and so the linear transformation of $W = span\{x_i, y_i \mid i = 1, \ldots, r\}$ to the dual S^* by $w \mapsto f(w, -)$ is onto. Let W_0 be the annihilator of S in W. Since $\dim S = r = \dim V/2$, $W_0 \leq S$. Clearly, $W_0^\perp = W_0 \perp span\{x_r, y_r\}$ and so $S = W_0 \perp (S \cap span\{x_r, y_r\})$, whence S is forced to contain a nonsingular vector.

(d) It suffices to show that $Q \perp I \cong D \perp I$, which is trivial. □

(2.24.1) Exercise. Let p be a prime number and x an element of order p in the orthogonal group $= O^\varepsilon(2n, q)$. Suppose that p divides $q^n - 1$, but not $q^k - 1$ for $k = 1, \ldots, n-1$. Show that x stabilizes a maximal totally singular subspace and that the Witt index is maximal.

(2.25) Witt's Theorem. *Let V be a vector space with a bilinear function f which is alternating, symmetric or Hermitian. Suppose that the subspaces U and W of V are isometric with respect to f, i.e., there is a linear isomorphism $h: U \to W$ such that $f(x,y) = f(h(x), h(y))$ for all $x, y \in U$. Then there exists an automorphism of V which preserves f and which restricts to h on U.*

The statement with "bilinear function" replaced by "quadratic form" is also true. See [Artin], p. 121. In the case of a quadratic form, we may not be able to choose the automorphism in the special orthogonal group; for instance, the maximal isotropic subspaces may fall into two orbits under the special orthogonal group.

(2.26) Notation. $(A\times B).> 2$ means the subgroup of $A.2\times B.2$ which contains the normal subgroup $A\times B$ and corresponds to the diagonal subgroup of the quotient. The extensions $A.2$ and $B.2$ are understood from context. We call $(A\times B).> 2$ the *simultaneous componentwise upward extension of* $A\times B$ *by* $\mathbb{Z}_2$. In the same vein, we call $2.<(A\times B)$ *the simultaneous componentwise downward extension of* $A\times B$ *by* $\mathbb{Z}_2$ if it is the quotient of some $2.A\times 2.B$ by the diagonal subgroup of order 2 in the extension kernel. We may replace $.>$ by $:>$ or $\cdot>$ to indicate split or nonsplit extensions, as appropriate.

(2.27) Presentations. We give several useful presentations for finite groups.

(2.27.1) $PSL(2,p)$; think of elements $\begin{pmatrix}1&0\\1&1\end{pmatrix}$ and $\begin{pmatrix}0&1\\-1&0\end{pmatrix}$:

$$s^p = t^2 = (st)^3 = 1 \text{ (for } p=2,3,5);$$

for p an odd prime, introduce v (corresponding to $\begin{pmatrix}\alpha&0\\0&\alpha^{-1}\end{pmatrix}$, where α is a primitive root modulo p) and the additional relation:

$$(tv)^2 = (s^\alpha tv)^3 = 1\,, \quad v^{-1}sv = s^{\alpha^2}\,.$$

See [CM], p. 94–95.

(2.27.2) The symmetric group of degree $n+1$ is presented by generators $r_i, i = 1,\ldots,n$, and relations $r_i^2 = 1$, $(r_ir_j)^2 = 1$ for $|i-j|\geq 2$ and $(r_ir_{i+1})^3 = 1$ for $i<n$. The elements r_i may correspond to the permutations $(i, i+1)$. See [Hu] or note that this is a special case of a presentation of a Coxeter group [CM], p. 63, [Car], Theorem 2.4.1. A covering group of the above, for $n+1\geq 4$, has generators z, q_i, $i=1,\ldots,n$, and relations $z^2=1$, $q_i^2 = z$, $(q_iq_{i+1})^3 = z$, $[q_i,q_j] = z$ for $i\neq j\pm 1$. Another covering group has generators $z, q_i, i=1,\ldots,n$, and relations $[z,q_i]=1$, $z^2=1$, $q_i^2=1$, $(q_iq_{i+1})^3=1$, $[q_i,q_j]=z$ for $i\neq j\pm1$. These groups are isomorphic only for $n+1=6$. [Schur11].

(2.27.3) The alternating group of degree n is presented by generators $s_1,\ldots,s_{n-2}$ and relations $s_1^3 = s_j^2 = (s_{j-1}s_j)^3 = 1$ for $2\leq j\leq n-2$, and $(s_is_j)^2 = 1$ for $1\leq i\leq j-2$, $j\leq n-2$. See [Hu] or [CM], p. 66. Define a group B_n by generators $z, b_1,\ldots,b_{n-2}$ and relations $Z^2 = 1$ $b_1^3 = z, (b_1b_2)^3 = z$, $(b_1b_k)^2 = z$ for $k>2$; for $i,j>1$: $b_i^2 = z$, $(b_ib_{i+1})^2 = z$, $[b_i,b_j] = z$ if $i\neq j\pm 1$.

(2.27.4) The Chevalley-Steinberg presentation of an untwisted group of Lie type of rank at least 2 over the field F.

Let Φ be an indecomposable root system of rank at least 2 and let $x_r(t)$ denote a symbol for each $(r,t) \in \Phi \times F$; if r is not a root, define $x_r(t) = 1$. Then, there are integers c_{ijrs} for each linearly independent set of roots r, s and each pair of positive integers i, j, so that, for all t, u in F,

(2.27.4.A) $x_r(t)\, x_r(u) = x_r(t+u)$;

(2.27.4.B) $[x_r(t), x_s(u)] = \prod_{i,j>0} x_{ir+js}\left(c_{ijrs} t^i u^j\right)$, product in order of increasing $i+j$;

(2.27.4.C) $h_r(t)\, h_r(u) = h_r(tu)$, for all $t, u \in F^\times$, where $n_r(t)$, for $t \in F^\times$, is defined by $n_r(t) := x_r(t)\, x_{-r}\left(-t^{-1}\right) x_r(t)$ and $h_r(t) := n_r(t)\, n_r(1)^{-1}$.

Steinberg proved that these generators and relations define a central extension of the Chevalley group of type Φ over F. For certain fields (e.g., finite) (2.27.4.C) follows from (2.27.4.A) and (2.27.4.B). See [St63, 67], [Car], Chapter 12. In case Φ has rank 1, we get a similar conclusion provided we replace (2.27.4.B) with

(2.27.4.B′) $n_r(t)\, x_r(u)\, n_r(t)^{-1} = x_{-r}\left(-t^2 u\right)$, for $t \neq 0$, u in F.

For generators and relations for twisted groups, see [St63, 67, 81], [Gr73b]. The group so presented is called the *Steinberg group* or *the universal group of type Φ over F*. It is often a covering group of the associated Chevalley group. More precisely, when the field is finite, the kernel of the epimorphism (2.12.5) is a p-group and, with finitely many exceptions, is an isomorphism (2.14).

(2.28) The Spin Trick and the Permutation Trick. This goes back to [Schur11] and was used by Steinberg, Griess and others. There is a double cover $\varphi: Spin(n, \mathbb{K}) \to SO(n, \mathbb{K})$, for a field $\mathbb{K}$ of characteristic not 2. Sometimes, if one embeds a group G in $SO(n, \mathbb{K})$, its preimage in $Spin(n, \mathbb{K})$ is a nonsplit central extension. Here are some sufficient conditions for this to be so.

(2.28.1) G has an involution whose spectrum has -1 with multiplicity 2 $(mod\, 4)$; Steinberg showed me this; it is easy to deduce from [Chev54].

(2.28.2) G has a faithful permutation representation π and an involution t in every subgroup of index 2 of G such that the number of letters moved by t^π is 4 $(mod\, 8)$. [Gr72a]

(2.28.3) [GaGa] contains character theoretic criteria for an orthogonal representation of a finite group to give a nonsplit extension.

The analysis of lifting subgroups of $SO(n, \mathbb{K})$ to $Spin(n, \mathbb{K})$ requires a study of spinors; see [Chev54], [GaGa]; certain highlights are discussed in [Gr91].

Chapter 3. Codes

Our purpose in this chapter is to review basic material about linear error correcting codes. The ones of most interest to us are over the fields of 2, 3 and 4 elements. These objects arise naturally when trying to describe structure and representations of almost simple groups. For further background on general coding theory, see [MacW-Sl].

We begin by studying linear algebra associated to the power set of a set. The results are needed later and they serve here to motivate several concepts in elementary coding theory.

(3.1) Definition. Let X be a set. The family of all subsets of X is called the *power set* and is denoted $P(X)$. It is a vector space over $\mathbb{F}_2$ via the addition operation $A + B := (A \backslash B) \cup (B \backslash A) = (A \cup B) \backslash (A \cap B)$, which is sometimes called the *symmetric difference operation* or the *Boolean sum.* Its zero-element is the empty set. Under the product operation $AB := A \cap B, P(X)$ becomes an algebra in which every element is idempotent, a so-called Boolean algebra. The family of subsets of even cardinality is denoted $PE(X)$, *the even sets.* It is a subspace, but not an ideal in general. When X is finite, this subspace has codimension 1. It is the kernel of the linear map $A \mapsto |A| \mod 2$. Note that $P(X)$ is isomorphic to $\mathbb{F}_2^X$ (= all set maps from X to $\mathbb{F}_2$), via the identification of a subset with its $\mathbb{F}_2$-valued characteristic function.

(3.2) Definition. Let V be a module over the commutative ring R. A function $Q : V \to R$ is called a *quadratic form* if and only if $Q(cv) = c^2 Q(v)$, for all $c \in R$ and $v \in V$ and if $f(v, w) := Q(v + w) - Q(v) - Q(w)$ is R-bilinear. We call f *the bilinear function associated to* Q. Write rad $V, rad(f)$ or $rad(V, f)$ for $\{x \in V \mid f(x, V) = 0\}$ and $rad(Q)$ or $rad_Q V$ for $\{x \in rad V \mid Qx = 0\}$. If the $rad_Q(V, f)$ is the 0-submodule, we say that the form Q is *nondegenerate*; otherwise, *degenerate.* If $rad(V, f) = 0$, we say that Q is *nondefective* ([Die], 33–34); otherwise *defective.* See also [Artin].

Obviously, f is symmetric. If $2R = 0$, f is also alternating. We remark that $rad V \neq 0$ and $rad_Q V = 0$ simultaneously is possible; see (3.4).

A vector is *singular* or *nonsingular* according to whether its value under the quadratic form in zero or nonzero. A subspace is *nonsingular* if the restriction to it of the bilinear form is nondegenerate and it is *totally singular* if the bilinear form restricts to the zero form; in the latter case, the quadratic form restricts to a homomorphism of abelian groups with image in $\{x \in R \mid 2x = 0\}$. A singular vector is also called *isotropic* and a nonsingular vector is also called *anisotropic* and these terms carry over naturally to points in projective space. See the books of Artin [Artin] and Dieudonne [Die] for generalities about classical groups and forms on vector spaces.

(3.3) Definition. Assume the notation of 3.1 and the finiteness of X. The map $Q : PE(X) \to \mathbb{F}_2$ defined by $A \mapsto \frac{1}{2}|A| (mod 2)$ and its associated bilinear form are called *the natural quadratic* and *bilinear forms on* $PE(X)$, respectively.

(3.4) Exercise. Verify that Q and f have the right properties and show that $f(A, B) = |A \cap B| \, mod 2$. The formula $g(A, B) = |A \cap B| \, (mod 2)$ makes sense on $P(X)$ and gives a bilinear function g on $P(X)$. Show that the radical of g is 0 and that the annihilator in $P(X)$ of $PE(X)$ is $\langle X \rangle$. Deduce that the radical of f is 0 or $\langle X \rangle$ according to whether $|X|$ is odd or even. Show that the quadratic form Q induces a well-defined quadratic form on $PE(X)/rad f$ (by taking its value on a coset to be the value of Q on a coset representative) if and only if $rad_Q(PE(X)) = rad(PE(X))$, i.e., $|X|$ is odd or $0 (mod 4)$.

(3.5) Exercise. Let Ω be a 7-set. The subspace $W := span\{v_1, v_2, v_3\}$ is singular, where $v_1 = (1111000)$, $v_2 = (1100110)$ and $v_3 = (1010101)$. Furthermore,any 3-dimensional singular subspace of minumum weight 4 in $PE\Omega$ may be transformed to W under coordinate permutations. Let $M := \mathbb{F}_2^3$ and $G := GL(M)$, the simple group of order 168. There is a natural action of G on the set of nonzero vectors of M, which we may identify with Ω. This gives an embedding of G into Σ_7 (hence into A_7, since G is simple). There is a G-epimorphism of modules $P\Omega \to M$ given by $A \mapsto \Sigma_{a \in A} a$. Let K be the kernel and let $L = K \cap PE\Omega$. Show that $dim\, K = 4$, $dim\, L = 3$ and that L is a totally singular subspace in which every nonzero vector is a 4-set. Identify geometrically (in terms of the vector space structure of M) the sets in L and in $K \backslash L$.

We are now ready to study basic concepts in coding theory.

(3.6) Definition. Let F be a field and $n \geq 0$ an integer. A *linear error-correcting code* is a subspace of F^n. For short, we use the term *code*. The integer n is called the *length* of the code.

(3.7) Definition. The *weight, or Hamming weight,* of a vector v in F^n is the number of nonzero coordinates; it is written $wt(v)$ or $wt\, v$. The *distance* between vectors x and y is $wt(x - y)$. The *minimum weight of a code* C is $\min\{wt\, x | x \in C, x \neq 0\}$.

(3.8) Notation. We say that a code is a $[n, d, w]$-*code* if its length is n, dimension is d and minimum weight is w. These integers are called the *code parameters.* We use the term *binary* for a code over $\mathbb{F}_2$ and *ternary* for a code over $\mathbb{F}_3$.

(3.9) Definitions. Suppose that F has an automorphism $\alpha \mapsto \bar{\alpha}$ of order 1 or 2. We define a Hermitian inner product $((x_i),(y_i)) := \Sigma_{i=1}^n x_i \bar{y}_i$ on F^n. If C is a subset of F^n, define $C^\perp := \{x \in F^n \mid (x, C) = 0\}$. If C is a code, we say that C is *weakly self orthogonal* if $C \subseteq C^\perp$ and is *self orthogonal* if $C = C^\perp$.

The term *self-dual* is sometimes used in place of self-orthogonal; however, in certain situations this term may be misleading since it is not true that a self-dual code with some automorphism group is self-dual as a module for that group. Consider the following definition.

(3.10) Definition. For a code $C \leq F^n$, the *cocode* is F^n/C.

In general, the cocode does not inherit a code structure since the weight function is not constant on cosets of C. Sometimes, we assign to a coset of C the minimum weight of its elements; see (5.9) for an example.

In case C is a weakly self-orthogonal code, the natural semilinear onto map $F^n \to C^* = Hom_F(C,F)$ factors through F^n/C. If C is self-orthogonal, the map $F^n/C \to C^*$ is an isomorphism.

(3.11) Exercise. Let F_0 be the fixed field under the field automorphism (3.11). Prove that the Hermitian form is nonsingular. When the Hermitian form $(\,,\,)$ is not bilinear (i.e., the field automorphism is nontrivial), one can define an F_0-bilinear form $\langle\,,\,\rangle$ on F^n by $\langle x,y\rangle = (x,y)+(y,x)$. Show that $\langle\,,\,\rangle$ is nonsingular. Conclude in this case that if F^n has a self orthogonal code C, n is even and $dim\, C = n/2$.

(3.12) Definitions. *The standard basis* of F^n is the set $e_1 := (1,0,\dots,0),\dots,e_n := (0,\dots,0,1)$. A *monomial transformation* is an F-linear map $F^n \to F^n$ such that $e_i \mapsto c_i e_{i\pi}$, for all i, where $c_i \in F^\times$ and π is some permutation. The *monomial group* of F^n is the set of all monomial transformations. This group permutes the *standard coordinate spaces* $\{E_i \mid i = 1,\dots,n\}$, where $E_i = Fe_i$. We write $Mon(n,F)$ for this group and note that it is a split extension of $Diag(n,F)$, *the diagonal group* for F^n (the set of invertible matrices preserving *each* E_i), by $Perm(n,F)$, *the group of coordinate permutations* (the group of invertible matrices preserving the set $\{e_i\}$), isomorphic to Σ_n. It is isomorphic to the wreath product, written $F^\times \wr \Sigma_n$ or $F^\times wr \Sigma_n$, with the natural action of Σ_n on n letters.

We have a natural action of $Aut(F)$, the group of field automorphisms of F, on F^n, by coordinatewise action. Let Γ be the group of transformations so obtained. The group Γ operates as semilinear transformations. We define *the extended monomial group* $Mon^*(n,F)$ to be the set of semilinear transformations which preserve the set of coordinate spaces. It is a semidirect product in these two ways: $Mon^*(n,F) = Mon(n,F){:}\Gamma = Diag(n,F){:}[Perm(n,F) \times \Gamma]$. Let τ be the field automorphism in (3.9). We define $Diag_1(n,F)$ as the set of matrices in $Diag(n,F)$ for which each diagonal matrix entry has norm 1, i.e., is in the kernel of $x \mapsto xx^\tau$; we also define $Mon_1^*(n,F) := Diag_1(n,F){:}Perm(n,F)\langle\tau\rangle$.

Note that every nonzero scalar has norm 1 in the cases of greatest interest to us: $F = \mathbb{F}_2, \mathbb{F}_3$ and, when $\tau \neq 1$, $F = \mathbb{F}_4$.

As is customary in group theory, when F is finite of order q, we replace F by q in above notations and write $Mon\,(n,q)$ for $Mon\,(n,F)$, etc.

(3.13) Definition. Let C be a code in F^n. *The automorphism group of the code* C is $Aut^*(C) := \{g \in Mon^*(n,F) \,|\, C^g = C\}$. We write $Aut\,(C)$ for $Aut^*(C) \cap Mon\,(n,F)$, and call this *the group of (strictly) linear automorphisms of the code.*

In general, the automorphism group of a code need not act faithfully on the code; see the discussion in (4.4).

(3.14) Examples. (a) Up to equivalence, there is a unique binary $[7,3,4]$-code, *the Hamming code.* Its group is a $GL(3,2)$-subgroup of $Perm(7,2) \cong \Sigma_6$; see (3.5).

(b) Up to equivalence, there is a unique binary $[8,4,4]$-code, called the *extended Hamming code.* Its group is the affine linear group $AGL(3,2) \cong 2^3{:}GL(3,2)$.

(c) Up to equivalence, there is a unique self-orthogonal ternary code of length 4; its parameters are $[4,2,3]$.

(3.15) Remark. The normal subgroup $Diag\,(n,F) \cap Aut\,(C)$ of $Aut^*(C)$ is in general not complemented. The groups of the hexacode (Section 4) and the tetracode (Section 7) are examples.

(3.16) Definition. We say that the codes C and D in F^n are *equivalent* iff there is $g \in \mathrm{Mon}^*(n,F)$ such that $C^g = D$. If such a g exists in $Mon\,(n,F)$, we say that the codes are *(strictly) linearly equivalent.*

Our definition of equivalence is not compatible with several of those given in [MacW-Sl]; see p. 24 and 39–40, lines 9–11.

Codes which are equivalent may not be isometric. However, orthogonality is preserved by $Mon_1^*(n,F)$. A code equivalent by such an element to a weakly self-orthogonal code is weakly self-orthogonal.

(3.17) Remark. If k is the number of codes equivalent to C in F^n, we have $k \cdot |Aut^*(C)| = |Mon^*(n,F)| = |F^\times|^n \cdot n! \cdot |Aut\,(F)|$.

(3.18) Definition. A *binary Golay code* is a $[24,12,8]$-code over $\mathbb{F}_2$.

In Section 5, we shall see that such a code exists, is unique up to equivalence and has a remarkable group of automorphisms.

(3.19) Definition *(Thickened codes).* Let C be a code in F^n and $k \geq 2$ a natural number. Let $I_j, j = 1,\dots,n$, be a partition of $\{1,\dots,kn\}$ into k-sets. Let $\varphi\colon F \to F^{kn}$ be the linear map which sends the j-th standard basis vector of F^n to the sum of standard basis vectors of F^{kn}, indexed by I_j. It has the property that $wt\,(x^\varphi) = k \cdot wt\,(x)$, for $x \in F^n$. A *k-thickened version of* C is a code C^φ.

(3.20) A formula for the number of isotropic subspaces, the Hermitian case.

Let $F \leq E$ be the fields of q and q^2 elements, respectively and let $x \mapsto \bar{x} = x^q$ be the nontrivial field automorphism. Let V be an n-dimensional vector space over E with a nondegerate hermetian form h. Define $I_n := \{x \in V | h(x,x) = 0\}$, $J_n := \{x \in V^{\#} | h(x,x) = 0\}$, $A_n := \{x \in V | h(x,x) \neq 0\}$ and let i_n, j_n, a_n be their respective cardinalities. We have

(3.20.1) $$i_n = j_n + 1 \text{ and } i_n + a_n = q^{2n} .$$

We seek a recurrence relation. Let $\{e_i | i = 1, \ldots, n\}$ be an orthonormal basis. Using coordinates, $v \in V$ may be written (c_i) and we have

(3.20.2) $$h(v,v) = \sum c_i \bar{c}_i .$$

Since the number of elements of E of norm $\nu \in F$ is 1 or $q+1$ as $\nu = 0$ or $\nu \neq 0$, we use the orthogonal decomposition $V = span\{e_i | i = 1, \ldots, n-1\} \perp E e_n$ and (3.20.2) to deduce that

(3.20.4) $$i_n = i_{n-1} + a_{n-1}(q+1) = q^{2n-2} + q a_{n-1} = q^{2n-2} + q^{2n-1} - q i_{n-1} ;$$

(3.20.5) $$a_n = q^{2n} - i_n = q^{2n} - q^{2n-2} - q a_{n-1} = q^{2n-2}\left(q^2 - 1\right) - q a_{n-1} .$$

Define $b_k := a_k/(q^2 - 1)$. Then (3.20.5) reads

(3.20.6) $$b_n + q b_{n-1} = q^{2n-2} .$$

A solution must satisfy $b_1 = 1$, $b_2 = q^2 - q$, $b_3 = q^4 - q^3 + q^2$, $b_4 = q^6 - q^5 + q^4 - q^3$, ... The formula

(3.20.7) $$b_n = q^{n-1}\left(q^{n-1} + (-1)^{n-1}\right) / (q+1)$$

solves the recurrence relation, as may be verified by induction.

(3.21) Exercise

(3.21.1) Verify all the steps in (3.20).

(3.21.2) Define $S(n,k)$ to be the number of k-dimensional totally singular subspaces in V.

(a) For which k is $S(n,k) \neq 0$?

(b) For each k, find a formula for $S(n,k)$.

Hint: Here is a possible approach. Let $x \neq 0$ be an isotropic vector. Then $x \in x^{\perp}$, which has codimension 1. If W is any complement to Ex in $x^{\perp}$, W is an $(n-2)$-dimensional vector space with a nonsingular hermetian form (the restriction of h). Then, the family of all $(k+1)$-dimensional isotropic subspaces of V containing x equals the number of k-dimensional isotropic subspaces of W.

(3.20.3) When $q = 2$, verify that $S(6,3) = 891$.

(3.20.4) How should this program be modified to work for nondegenerate symmetric bilinear forms over $\mathbb{F}_q$?

(3.22) Definition. The *pentacode* is a $[6,3,4]$ code over $\mathbb{F}_5$; it is spanned by

(3.22.1) $$(101023),\ (102310) \text{ and } (231010).$$

The *septacode* is a $[4,2,3]$ code over $\mathbb{F}_7$ spanned by

(3.22.2) $$(5111) \text{ and } (0124).$$

(3.23) Exercise. Show that the automorphsims of the pentacode and septacode are, respectively, $4 \times Sym_5$ and $3 \times SL(2,3)$. Determine the orbits of these groups on the set of codewords and these codes and cocodes as modules for the automorphism groups.

Chapter 4. The Hexacode

We establish basic properties of the hexacode, including existence and uniqueness. Mastery of the hexacode is essential to our treatment of the Golay code. We use the notations $0, 1, \omega$ and $\bar{\omega}$ for the elements of $\mathbb{F}_4$ with the obvious meanings for 0 and 1 and with $\omega^2 = \bar{\omega} = 1 + \omega, \bar{\omega}^2 = \omega = 1 + \bar{\omega}$ and $\omega\bar{\omega} = 1$. We often use the simple fact that $v \in \mathbb{F}_4^n$ is isotropic, i.e., $(v, v) = 0$, iff $wt\,(v)$ is even.

(4.1) Definition. A *hexacode* is a linear code over $\mathbb{F}_4$ of dimension 3, length 6 and minimum weight 4. The inner product on $\mathbb{F}_4^6$ is the usual Hermitian dot product $(x_i)\,(y_i) = \sum_{i=1}^{6} x_i \bar{y}_i$.

We write $\mathcal{H}$ for the linear span of the following elements of $\mathbb{F}_4^6$:

$$\begin{gathered}(\omega\bar{\omega}|\omega\bar{\omega}|\omega\bar{\omega})\\(\bar{\omega}\omega|\bar{\omega}\omega|\omega\bar{\omega})\\(\omega\bar{\omega}|\bar{\omega}\omega|\bar{\omega}\omega)\\(\bar{\omega}\omega|\omega\bar{\omega}|\bar{\omega}\omega)\ .\end{gathered}$$

Notice that $\mathcal{H}$ is 3-dimensional, self orthogonal, and that any three of these four vectors forms a basis. We call $\mathcal{H}$ the *standard hexacode.*

We write the six coordinates in blocks of two with adjacent blocks separated by vertical lines. The reason for this notation will become clear.

(4.2) Definition *(The subgroup $Z \times S$ of $Aut\,(\mathcal{H})$).* The group $Z \cong \mathbb{Z}_3$ operates as multiplication by powers of ω and $S \cong \Sigma_4$ operates in the following way as a group of natural coordinate permutations: there is a semidirect product decomposition $S = VT$, $V = O_2(S) \cong \mathbb{Z}_2 \times \mathbb{Z}_2, T \cong \Sigma_3$ such that elements of the group V preserve each block while switching coordinates within evenly many blocks and elements of the group T permute the three blocks while preserving the order of coordinates within each block.

It is straightforward to check by looking at the spanning set for $\mathcal{H}$ that the group generated by Z, V and T is indeed in the group of the code and has the claimed structure. A straightforward calculation shows that T normalizes V and that $VT \cong \Sigma_4$. So, we have a homomorphism of $S = VT$ into $Perm\,(6, 4)$.

(4.2.1) Exercise. Give details for the following alternate verification that $S = VT$ has a permutation representation with the properties listed in the above paragraph. Partition a 6-set X into three 2-sets X_1, X_2 and X_3. Let H be the subgroup of $Sym(X)$ preserving this partition. Show that the natural map π of H to $Sym(\{X_1, X_2, X_3\})$ is onto and that the kernel is isomorphic to $\mathbb{Z}_2^3$. Define $L = H \cap Alt(\{X_1, X_2, X_3\})$. Show that $\pi|_L$ is onto, that $L \cong \Sigma_4$ and that this permutation representation of Σ_4 has all the properties of the putative permutation group S of the preceeding paragraph. Notice that S is transitive on X. □

We now describe the vectors in $\mathcal{H}$ by giving representatives for the orbits of ZS. The proof is not given and is easy with all the numerical information given. To check that a vector is in $\mathcal{H} = \mathcal{H}^\perp$, it suffices to check that $(v, x) = 0$, for all x in a spanning set, since $\mathcal{H}$ is self-orthogonal. To verify that the stated subgroups are the full stabilizers, just find sufficiently many images of the orbit representative under ZS.

Table 4.1. The orbits of ZS on $\mathcal{H}$

Orbit representative	Stabilizer in ZS	Orbit length
$(0\,0\vert 0\,0\vert 0\,0)$	ZS	1
$(0\,0\vert 1\,1\vert 1\,1)$	Dih_8	9
$(1\,1\vert \omega\,\omega\vert \bar\omega\,\bar\omega)$	$V\langle tz\rangle \cong Alt_4$	6
$(0\,1\vert 0\,1\vert \omega\,\bar\omega)$	$\Sigma_2\ (\leq T)$	36
$(\omega\,\bar\omega\vert \omega\,\bar\omega\vert \omega\,\bar\omega)$	$T \cong \Sigma_3$	12

($t \in T$ is a 3-cycle and $z \in Z$ is multiplication by ω; see (4.2))

We now prove uniqueness results for self-orthogonal codes in certain $\mathbb{F}_4^n$. By (3.11), n is even.

(4.3) Lemma. *Let C be a self-orthogonal code in $\mathbb{F}_4^4$. Then, up to equivalence (even to strict linear equivalence) in the sense of* (3.14), *C is the span of* $(1, 1, 0, 0)$ *and* $(0, 0, 1, 1)$.

Proof (see (3.15)). If C contains a vector v of weight 2, we may assume that $v = (1100)$. Since $C^\perp$ is contained in $v^\perp = span\,\{v, (00**)\}$, we are done. Now just observe that C must contain a vector of weight 2: for $dim\, C = 2$ implies that C contains two independent 4-tuples, and some linear combination of them is nonzero but has a 0-coordinate, hence two such. □

(4.4) Proposition. *Let C be a self-orthogonal code in $\mathbb{F}_4^6$. Then, up to linear equivalence in the sense of (3.15), C is one of the following two codes:*

$C_1 := span\,\{(1,1,0,0,0,0), (0,0,1,1,0,0), (0,0,0,0,1,1)\}$; or
$\mathcal{H} := span\,\{(\omega\,\bar\omega\vert\omega\,\bar\omega\vert\omega\,\bar\omega), (\bar\omega\,\omega\vert\bar\omega\,\omega\vert\omega\,\bar\omega), (\omega\,\bar\omega\vert\bar\omega\,\omega\vert\bar\omega\,\omega)\}$, the standard hexacode.

Proof: Of course, $dim\, C = 3$; see (3.10). Suppose C contains a vector v of weight 2. Without loss, $v = (110000)$. Since $C^{\perp}$ is contained in $v^{\perp} = span\, \{v, (00{*}{*}{*}{*})\}$, we use (4.3) to get $C = C_1$.

Now assume C contains no vector of weight 2. There are vectors of weight 4; for otherwise there must be two independent vectors of weight 6 and some nonzero linear combination of them has a zero coordinate. Let v be a vector in C of weight 4. We may assume that $v = (111100)$.

For a vector $w \in \mathbb{F}_4^6$, let w_i denote the i-th coordinate. We may assume that $v_5 = v_6 = 0$. Let v, x, and y be a basis of C. Choose x and y so that $x_1 = y_1 = 0$ and $y_2 = 0$. Since C is self-orthogonal, weights are even, so that x and y have weight 4. Since $0 = (v, y) = y_3 + y_4$, $y_3 = y_4$. Applying a diagonal transformation, we may arrange $y = (001111)$ and still have $x_1 = 0$. We arrange $x_3 = 0$ by replacing x with $x - x_3 y$. Replacing x by $x_2^{-1}x$, we get $x_2 = 1$. Since $0 = (x, v) = 1 + x_4$, $x_4 = 1$. Similarly, $0 = (x, y)$ gives $x_5 + x_6 = 1$. Since x_5 and x_6 are nonzero scalars, they are ω and $\bar{\omega}$., in some order. Reindexing if necessary, we have $x = (0101\omega\bar{\omega})$. Therefore, $C = span\, \{v, x, y\} = \mathcal{H}$. □

We now determine $Aut\, C$ and $Aut^*\,(C)$, for a self-orthogonal code in $\mathbb{F}_4^6$. See the definitions in Section 3.

Note that the standard field automorphism fixes C_1 but not necessarily a code equivalent to C_1, and that $\mathcal{H}$ is not stable under the standard field automorphism. Note also that $Aut^*\,(C_1) \cong \left(\left(\mathbb{F}_4^{\times} \times \Sigma_2\right) wr \Sigma_3\right){:}2 \cong [\mathbb{Z}_6 wr \Sigma_3]{:}2$ and that the kernel of the action is isomorphic to $\Sigma_2 \times \Sigma_2 \times \Sigma_2$.

(4.5) Proposition. *In the notation of* (4.4),

(i) *$Aut\,(C_1)$ is isomorphic to $\mathbb{Z}_6 wr \Sigma_3$, where the wreath product is taken with respect to the natural action of Σ_3 on* 3 *letters; its intersection with the diagonal group is isomorphic to $\mathbb{Z}_3^3$ and $Aut^*\,(C_1)$ contains the field automorphism; the kernel of its action on C_1 is a group of permutation matrices isomorphic to $\mathbb{Z}_2^3$.*

(ii) *$Aut^*\,(\mathcal{H})$ is the unique nonsplit extension $3 \cdot \Sigma_6$* (2.15.4)*; its intersection with the group of diagonal matrices is the group of scalar matrices, the subgroup $Aut\,(\mathcal{H})$ acting as $\mathbb{F}_4$-linear transformations is the group $3^{\bullet}Alt_6$ of index* 2, *and the group induced on the set of* 6 1*-dimensional subspaces spanned by the standard basis vectors of shape $\left(1^1 0^5\right)$ is Σ_6.*

Proof. (i) This is obvious once we observe that the only vectors of weight 2 in C_1 are the scalar multiples of the three standard basis elements.

(ii) Throughout this part of the proof, it would be helpful to know standard information about the theory of central extensions of Alt_n and Σ_n. A good reference for this is [Hu], V.23–25 and pp. 652–654. See also Section 2. Note that the field automorphism is not in $Aut^*\,(\mathcal{H})$ since the image of $(01|01|\omega\bar{\omega})$ is $(01|01|\bar{\omega}\omega)$, which is not in $\mathcal{H}$.

The group ZS is contained in $Aut\,(\mathcal{H})$. On X, the set of 1-dimensional subspaces spanned by the standard basis vectors of $\mathbb{F}_4^6$, ZS induces a transitive but imprimitive Σ_4-subgroup of Σ_X which commutes with a permutation with cycle shape 2^3. We display a semilinear transformation, α, of $\mathbb{F}_4^6$ which preserves the code $\mathcal{H}$, namely $\alpha = A\varphi$, where φ is the nontrivial field automorphism on coordinates and A is a linear transformation with matrix

$$\begin{pmatrix} 1&0&0&0&0&0 \\ 0&0&1&0&0&0 \\ 0&1&0&0&0&0 \\ 0&0&0&1&0&0 \\ 0&0&0&0&\bar{\omega}&0 \\ 0&0&0&0&0&\omega \end{pmatrix}.$$

Recall that groups act on the right, so α is computed by first applying A to a row vector, then applying φ, One can check directly that $\alpha^2 = 1$ and that α preserves the span of the vectors $(\omega\,\bar{\omega}|\omega\,\bar{\omega}|\omega\,\bar{\omega})$, $(\bar{\omega}\,\omega|\bar{\omega}\,\omega|\omega\,\bar{\omega})$ and $(\omega\,\bar{\omega}|\bar{\omega}\,\omega|\bar{\omega}\,\omega)$ (e.g., just check that the image of each under α is in $\mathcal{H}^\perp = \mathcal{H}$). It is then an easy exercise to prove that the action of α on the set X is the transposition (23) in the natural indexing and that this transposition and the above Σ_4-subgroup generate all of Σ_X.

Under the action of the group $G := \langle ZS, \alpha\rangle$ on $Diag\,(6,4)$ by conjugation, $Diag\,(6,4)$ has a unique composition series with Löwey factors of dimension 1,4 and 1; see (1.8) on the $\mathbb{F}_3\,Alt_6$-permutation module structure (note that $Diag(6,4)$ is not a permutation module for $\mathbb{F}_3\Sigma_6$; it is the tensor product of the permutation module with the sign representation since semilinear maps are involved here; thus, $Diag\,(6,4)$ and the permutation module for $\mathbb{F}_3\Sigma_6$ have identical lattice of submodules). The codimension 1 submodule is the set of diagonal matrices of determinant 1. It follows that if G has a nonscalar diagonal transformation, it must have all diagonal transformations of determinant 1; in particular, it has one, say g, with exactly two non-1 entries, ω and $\bar{\omega}$. Then $(\omega\bar{\omega}|\omega\bar{\omega}|\omega\bar{\omega})\,(g-1)$ would be a forbidden vector of weight 2. Therefore, $G \cap Diag\,(6,4) = Z$; let z generate Z. We now know that G is part of an exact sequence of one of the following forms: $1 \to Z \to G \to \Sigma_6 \to 1$ or $1 \to \Sigma_3 \to G \to \Sigma_6 \to 1$.

To prove that G has the form $3{\cdot}\Sigma_6$, it suffices to show that G contains a nonsplit extension G_1 of shape $3{\cdot}\Sigma_6$ (there is a unique such group; see (2.15.4)); for then, if $Aut^*(\mathcal{H})$ were larger than G_1 (i.e., if the second of the two exact sequences were the right one), it would contain an element t which is central $mod\,O_3\,(G)$ and so acts trivially on $G/O_3\,(G)$; such an element must lie in $Diag\,(6,4)\,\langle\varphi\rangle$, whence $|t| = 2$, since φ inverts $Diag\,(6,4)$, and t is semilinear and not linear, whence t is nontrivial on Z; the Frattini argument proves that $G = \langle Z,t\rangle C_G\,(t) = ZC_G\,(t)$, a semidirect product. Since $C_G\,(t)\,/\langle t\rangle \cong \Sigma_6$ and $|C_G\,(t) : C_G\,(\langle Z,t\rangle)\,| = 2$, we get $C_G\,(t) \cong \Sigma_6 \times \mathbb{Z}_2$ and that G must have Sylow 3-group $\mathbb{Z}_3^3$, since it is a product of Z with a Sylow 3-group of $C_G\,(t)$. This is impossible, as we shall see in the next paragraph.

We give several proofs that G is a nonsplit extension of Σ_6 by $\mathbb{Z}_3$. We first produce a nonabelian subgroup of G of order 27. A direct calculation shows that such a subgroup is generated by β and γ, where $\beta = [v,\alpha]$, $\gamma = \beta^{gh}$, $v \in V^\#$ is trivial on the second block, and $g \in R$ interchanges blocks 1 and 3, and $h \in V$ interchanges entries in blocks 1 and 2; thus, under the natural map of G onto

Σ_6, β goes to (123) and γ goes to (456). This calculation will be more transparent after we study the Golay code: for β and γ, just take the permutations UP8 and UP13 (see (5.39)). A second method is to use the fact that G must act faithfully on $\mathcal{H} \cong \mathbb{F}_4^3$, whence the subgroup $3.Alt_6$ of index 2 in G is embedded in $SL(3,4)$ as a subgroup of index prime to 3. Since $SL(3,4)$ has nonabelian Sylow 3-subgroup of order 27, so does G. □

(4.6) Exercises

(4.6.1) *(Alternative definitions of the standard hexacode).* For an element $x = (x_1, x_2, x_3, x_4, x_5, x_6)$, define $s_i := x_{2i-1} + x_{2i}$ for $i = 1, 2, 3$. When $s_1 = s_2 = s_3$, we say that x *has a slope* and we define the *slope* of x to be their common value. Show that $\mathcal{H}$ is defined by the requirement that vectors have a common slope, s, and satisfy the following additional linear condition, called *hexacode balance,* for i, j and k in different coordinate blocks:

$$x_i + x_j + x_k = \begin{cases} \omega s & \text{if evenly many of } i, j, k \text{ are even} \\ \bar{\omega} s & \text{otherwise.} \end{cases}$$

The right side may be expressed by the formula $\omega^{(-1)^{i+j+k+1}} s$.

Another balance concept plays an important role later; see (5.13), ff.

(4.6.2)

(4.6.2.1) Show that the action of $Aut^*(\mathcal{H})$ on $\mathcal{H}$ has orbits with representatives $(00|00|00)$, $(\omega\,\bar{\omega}|\omega\,\bar{\omega}|\omega\,\bar{\omega})$ and $(00|1\,1|1\,1)$ and respective lengths 1, 18 and 45, i.e., two hexacode words are in the same orbit iff they have the same weight. Determine all systems of imprimitivity for these two permutation representations. *Hint*: consider what α does to the orbit representatives from Table 4.1. Notice that the action on the six 1-dimensional spaces of vectors of weight 6 give a permutation representation of $Aut^*(\mathcal{H})$ which is inequivalent to the action on the six coordinate spaces in $\mathbb{F}_4^6$. Prove that the stabilizer of one such space is isomorphic to $[3 \times Alt_5]:2 \cong 3{:}\Sigma_5$.

(4.6.2.2) Show that in $G = \Sigma_6$, there are two conjugacy classes of maximal subgroups of order divisible by 5 and that both are isomorphic to Σ_5. *Hint:* First, prove that a maximal solvable subgroup of G of order divisible by 5 is isomorphic to 5:4, a Frobenius group of order 20 (results in Section 1 should be helpful here). Next, prove that a nonsolvable subgroup H of G' of order divisible by 5 satisfies $O_{5'}(H) = 1$ and contains a dihedral subgroup D of order 10. It is also not 2-nilpotent, so has order divisible by 12. Thus, H has order divisible by 60 and index dividing 6 in G', which is the simple group Alt_6. This forces $H \cong Alt_5$. If t is an involution in D, $C_{G'}(t)$ is a Sylow 2-group, dihedural of order 8. Show that a choice of noncentral involution u from $C_{G'}(t)$ leads to such an H as $\langle D, u\rangle$; there are essentially two such choices. Deduce the situation with G from that of G'.

(4.6.2.3) Describe the stabilizers in $Aut^*(\mathcal{H})$ of vectors of weight 4 and weight 6 and their actions on the coordinate spaces. Describe the submodule structure of $\mathcal{H}$ with respect to these stabilizers.

(4.6.3) Find $\mathbb{F}_2$-subspaces of largest possible dimension in $\mathcal{H}$, all of whose nonzero vectors have weight 4 (weight 6, respectively). Find the stabilizers in $Aut^*(\mathcal{H})$.

(4.6.4) *(Error correcting properties of the hexacode).* Let $I = \{1, 2, \ldots, 6\}$ index the 6-tuples. Let $J \subseteq I$ and $f\colon J \to \mathbb{F}_4$. Show that if $f : J \to \mathbb{F}_4$, there is a unique $g : I \to \mathbb{F}_4$ which gives a hexacode word and $g|_J = f$ (*"three given coordinates are part of a unique hexacode word"*). Show that if J is a 5-set, there is a unique $g : I \to \mathbb{F}_4$ giving a hexacode word and such that $g|_J$ and f disagree at at most one index (*"5 given coordinates, possibly after correcting at most one index, are part of a unique hexacode word."*). Hint: Completion: the linear transformation $\mathcal{H} \to \mathbb{F}_4^3$ gotten from projecting to the sum of three coordinate spaces must be monic since the minimum weight in $\mathcal{H}$ is 4. Correction: say the omitted coordinate is the sixth; let ψ be the projection of $\mathcal{H}$ to Y, the sum of the first five coordinate spaces; then $Im(\psi)$ is a $[5, 3, 3]$ code and the sets $S(c) := \{x \in Y \mid wt(x - c) \leq 1\}$ of cardinality $1 + 3 \cdot 5 = 16$ are, for $c \in Im(\psi)$, pairwise disjoint and so partition Y (thus $Im(\psi)$ is a *perfect* 1-*error correcting code*). Thus, ψ is an injection, which is what we want.

(4.6.5) Use (4.6.1) to carry out the error corrrecting procedure of (4.6.4). Completion: either the sum of two of the given coordinates is the slope or else we can get it from hexacode balance, (4.6.1). Correction: say, for ease of notation, that coordinates 1 through 5 are given; if $s_1 = s_2$, then $x_1, \ldots, x_4$ are correct and we get the slope and use (4.6.1) to get x_5 and x_6; if $s_1 \neq s_2, x_5$ is correct and $s = s_1$ or s_2, whence use of hexacode balance detects which of the first two blocks has the incorrect coordinate.

(4.6.6) *The "evenly many accents graves" rule.* This is a very useful observation: within each block $|ab|$, place an accent, ′ or ˋ, according to the following law: *accent aigu* (′) if $a = 0$, $a = b$ or $a\omega = b$; otherwise *accent grave* (ˋ). A hexacode word must have evenly many *accents graves* (the proof is an exercise, using Table 4.1 and the group ZS of Section 4). For example, the rule tells us to complete $(1\omega|\bar{\omega}0|**)$ to $(1\omega|\bar{\omega}0|\bar{\omega}0)$ rather than $(1\omega|\bar{\omega}0|0\bar{\omega})$. It is useful to speak of a relation (*not* a partial ordering) on $\mathbb{F}_4$ defined by $x < y$ iff the block $|xy|$ gets an accent aigu; see (5.28). □

(4.6.7) *Alternate description of the orbits in Table 1.* The orbits are the five equivalence classes of $\mathcal{H}$ under the equivalence relation $x \sim y$ iff

(4.6.7.1) $wt(x) = wt(y)$; and

(4.6.7.2) the common slopes of x and y are both 0 or both nonzero.

I thank Ernie Shult for this observation.

Chapter 5. The Golay Code

We begin by exploring the relationship between the binary Golay code (see (3.17)) and the Steiner system with parameters $(5, 8, 24)$.

(5.1) Definition. A *Steiner system* with parameters (a, b, c) is a family $\mathcal{S}$ of subsets of some set Ω of c elements such that (i) $A \in \mathcal{S}$ implies that $|A| = b$; (ii) for any a-set F in Ω, there is a unique $A \in \mathcal{S}$ such that $F \subseteq A$.

From now through (5.11), we assume that one exists, and discuss its properties. Starting with (5.12), we head for an existence result (5.16) for such a Steiner system and a uniqueness result (5.29).

(5.1.1) Notation. Let $\mathcal{S}$ be a Steiner system with parameters $(5, 8, 24)$. We call elements of $\mathcal{S}$ *octads*, a term used in the article [Todd].

At once, we see that $\mathcal{S}$ has $\binom{24}{5}/\binom{8}{5} = 759$ octads; more on this in (5.2).
We may regard $\mathcal{S}$ as a subset of $P\Omega$ and study $\mathcal{S}$ with respect to the ideas in Chapter 3 on $P\Omega$.

(5.2) Lemma. *Let N_A be the number of octads containing the subset A of Ω. Then $N_A = \binom{24-k}{5-k}/\binom{8-k}{5-k}$, if $k = |A| \leq 5$, and $N_A = 0$ or 1 otherwise.*

Proof. Suppose $k \leq 5$. We count pairs $(A, \mathcal{O})$, where A is a k-set and $\mathcal{O}$ is an octad containing A. We fill out A to a 5-set, which is contained in a unique octad. There are $\binom{24-k}{5-k}$ ways to do this. The same octad is achieved in this manner by exactly $\binom{8-k}{5-k}$ completions of A to a 5-set in $\mathcal{O}$. It follows that N_A is independent of the choice of k-set, A. The statement for $k \geq 5$ is obvious. □

We use the notation $(-1)^S$ for $(-1)^{|S|}$, when S is a finite set.

(5.3) Lemma. *Let $A \subseteq B \subseteq \Omega$. Let $N_{B,A}$ be the number of octads which meet B in A exactly (so that $N_{A,A} = N_A$; see (5.2)). Let y be any point of $\Omega \backslash B$, B' denote $B \cup \{y\}$ and let A' denote $A \cup \{x\}$, where x is any point of $B' - A$. Then*
(i) *(if $x = y$ or B' lies in an octad) $N_{B,A} = N_{B',A} + N_{B',A'}$;*
(ii) *(in general) $N_{B,A} = \sum_{A \subseteq S \subseteq B} (-1)^{S \backslash A} N_S$.*

Proof. If $x = y$, a picture suffices to prove (i). To prove (ii), use induction on $|B - A|$ and the case $x = y$ of (i). Namely, let priming denote the set obtained by adding a common point, say z, to A and B. We assume the result for $|B\backslash A| = |B'\backslash A'|$. Then, $\sum_{A \subseteq S \subseteq B'} (-1)^{S\backslash A} N_S = \sum_{A \subseteq S \subseteq B} (-1)^{S\backslash A} N_S + \sum_{A' \subseteq S \subseteq B'} (-1)^{S\backslash A} N_S = N_{B,A} - N_{B',A'}$, which implies the result for $|B'\backslash A|$, using the $x = y$ case of (i).

Finally, observe that, when $A \subseteq S \subseteq B \subseteq \mathcal{O}$, N_S depends only on $|S|$ and so (ii) and (5.2) imply that $N_{B,A}$ depends only on the cardinalities of A and B, so that (i) holds even for $x \neq y$. □

(5.4) The Leech Triangle. In the table below, which appeared in [Co69a, 71], the $(j+1)$-th entry in the $(i+1)$-th row is $N_{B,A}$, where $A \subseteq B$, B is an i-set and A is a j-set. It is computed from (5.2) and (5.3). Note that the N_A run along the right slope of the triangle.

Table 5.1. The Octad Triangle

$N_{i,j}$ is the common value of $N_{B,A}$ when the value is dependent only on the cardinalities of A and B.

759
506 253
330 176 77
210 120 56 21
130 80 40 16 5
78 52 28 12 4 1
46 32 20 8 4 0 1
30 16 16 4 4 0 0 1
30 0 16 0 4 0 0 0 1

(5.5) Lemma. *The subspace $\mathcal{G}$ of $P\Omega$ generated by $\mathcal{S}$ is totally singular, hence of dimension at most* 12.

Proof. If $A, B \in \mathcal{S}$, $|A \cap B|$ is even; this follows from examination of row 9 of the Leech table. Thus, $\mathcal{G}$ is a totally singular subspace. Since this bilinear form is nonsingular, we get $dim\, \mathcal{G} \leq |\Omega|/2 = 12$. □

(5.6) Definition. A *dodecad* is a 12-set in $\mathcal{G}$.

(5.7) Lemma. *Dodecads exist. In fact, there are at least* 2576 *dodecads which are writeable as a sum of two octads.*

Proof. The Octad Triangle tells us that $N_{8,2} = 16$. Therefore, dodecads exist since $\mathcal{D} = \mathcal{O} + \mathcal{O}'$ is one, where $\mathcal{O}$ and $\mathcal{O}'$ are octads which meet in a 2-set.

Let $\mathcal{A}$ be the set of dodecads which are expressible as the sum of two octads and let $\mathcal{B}$ be all ordered pairs of octads which meet in a 2-set. There is a natural map $f: \mathcal{B} \to \mathcal{A}$, $(\mathcal{O}, \mathcal{O}') \mapsto \mathcal{O} + \mathcal{O}'$. Define $\mathcal{B}_{\mathcal{O}} = \{(\mathcal{O}_1, \mathcal{O}_2) \in \mathcal{B} \mid \mathcal{O}_1 = \mathcal{O}\}$. Then $\mathcal{B}$ has $|\mathcal{S}|.|\mathcal{B}_{\mathcal{O}}| = 759.\binom{8}{2}16 = 340032$ elements.

Given $\mathcal{D} \in \mathcal{A}$, we consider the set of expressions $\mathcal{D} = \mathcal{O}_1 + \mathcal{O}_2$. Such an expression leads to a 6-set in $\mathcal{D}$ which is contained in an octad, namely $\mathcal{D} \cap \mathcal{O}_1$; let n be the number of octads contained in $\mathcal{D}$ (this will turn out to be 0, by (5.8)). We have $|f^{-1}(\mathcal{D})| = \left[\binom{12}{5} - \binom{8}{5}n\right]/6 \leq 132$. Therefore, $|\mathcal{A}| \geq 340032/132 = 2576$. □

(5.8) Theorem. *$dim\, \mathcal{G} = 12$ and the weight distribution of $\mathcal{G}$ is the following:*

weight	number of vectors
0	1
8	759
12	2576
16	759
24	1

Furthermore, if a set is in $\mathcal{G}$, so is its complement. No octad is contained in a dodecad. Every dodecad is the sum of two octads in 132 *different ways. The code $\mathcal{G}$ is self-orthogonal and has parameters* $[24, 12, 8]$.

Proof. Let $d = dim\, \mathcal{G}$. Since $\mathcal{S}$ has at least 2576 dodecads, $2^d > 2576$, whence $d = 12$ by (5.5). Since $d = 12$ and $\mathcal{G}$ is a singular subspace, it is self-orthogonal and so every element is even. The element Ω in $P\Omega$ annihilates every even set, hence $\Omega \in \mathcal{G}$. Since our lower bounds for the weights 0, 8, 12, 16 and 24 sum to $2^{12} = 4096$, the lower bounds equal those respective weights. If a dodecad contains an octad their sum is a 4-set contained in $\mathcal{G}$, a contradiction (so, in the proof of (5.7), $n = 0$). □

We now study the structure of $P\Omega/\mathcal{G}$, the Golay cocode; see (3.10).

(5.9) Theorem. *Let G denote a* $[24, 12, w]$ *binary code, with $w \geq 8$. Then, $w = 8$, and every coset of G in $P\Omega$ may be given a uniquely defined weight n from 0 to 4 according to whether it contains an n-set. A coset of weight n contains exactly 1 vector of weight n if $n = 0, 1, 2$ or 3 and 6 of weight 4 if $n = 4$. We have* $\binom{24}{0} + \binom{24}{1} + \binom{24}{2} + \binom{24}{3} + \frac{1}{6}\binom{24}{4} = 4096$.

Proof. Take a coset C of G. If C contains distinct vectors x, y of weights m, $n \leq 4$, then $x+y$ is nonzero of weight at most $m+n$. But the weight is at least 8, whence $m = n = 4$ and $x+y$ is an 8-set in G. Thus, a lower bound for the number of cosets is $\binom{24}{0} + \binom{24}{1} + \binom{24}{2} + \binom{24}{3} + \frac{1}{6}\binom{24}{4}$. An upper bound is $|P\Omega/G| = 2^{12}$, which equals the lower bound, whence all our inequalities are equalities and we have accounted for all the cosets. □

(5.10) Definition. A *sextet* is a partition of Ω into six 4-sets, the union of any two of which is an octad.

Thus, a sextet is simply all 4-sets in a coset of $\mathcal{G}$ of weight 4. The number of sextets is $\frac{1}{6}\binom{24}{4} = 10626/6 = 1771 = 7 \cdot 11 \cdot 23$. A 4-set is part of a unique sextet.

(5.11) Proposition. *There is a bijection between Steiner systems on Ω with parameters $(5, 8, 24)$ and Golay codes in $P\Omega$ (i.e., binary $[24, 12, 8]$-codes), obtained in the following way: the linear span of a Steiner system is a Golay code and the set of vectors of weight 8 in a Golay code is a Steiner system.*

Proof. We showed that a Steiner system determines a Golay code in (5.8). Let $\mathcal{G}$ be a Golay code and $\mathcal{S}$ the set of weight 8 vectors in it. We argue that $\mathcal{S}$ is a Steiner system. Then (5.9) implies that each coset has one vector of weight 0, 1, 2, 3 or 6 of weight 4 and that two such distinct weights may not occur in a given coset. Therefore, if F is a 5-set, the coset $F + \mathcal{G}$ must have minimum weight $n \leq 4$, say at $F + N$, $N \in \mathcal{G}$. If $n = 4$, let Ξ be the set of 4-sets in the coset $F + \mathcal{G}$; Ξ partitions Ω. Since $n = 4$, no part of Ξ is contained in F. Let $U \in \Xi$ contain a point of F. Then $U + F \in \mathcal{G}$ contains between 1 and 7, points, a contradiction. Therefore, $n \leq 3$ and so $n = 3$ and F is indeed contained in an 8-set $F + N = F \cup N$ of $\mathcal{G}$. Observe that such an 8-set in $\mathcal{G}$ is unique. It follows that the family of all 8-sets of $\mathcal{G}$ is a Steiner system with parameters $(5, 8, 24)$. Also, observe that by (5.8), the linear span of this family is a Golay code, which is in our $\mathcal{G}$, hence equals it since both have dimension 12. □

Now we turn to the existence question. Until further notice, we fix the following notation.

(5.12) Notation. Ω is a 24-set; Ξ is an ordered partition of Ω into six 4-sets; l is a function $\Omega \to \mathbb{F}_4$, called the *scalar labeling*, whose restriction to each set in Ξ is a set isomorphism; the *associated picture* is a picture of Ω with the parts indicated and the values of l placed on the points. Such a picture is *standard* if it is a 4×6 arrangement of Ω satisfying these conditions:

	1	2	3	4	5	6
0	·	·	·	·	·	·
1	·	·	·	·	·	·
ω	·	·	·	·	·	·
$\bar{\omega}$	·	·	·	·	·	·

The function l is constant on rows, taking the respective values $0, 1, \omega$ and $\bar{\omega}$ (from top to bottom), and the columns $K_1, \ldots, K_6$ are the members of $\Xi_{\bullet}$. We define the row R_c as $l^{-1}(c)$, for $c \in \mathbb{F}_4$.

We assume that Ξ and l are fixed and we work with the standard picture unless stated otherwise (e.g. in (5.36)).

The scalar labeling map l gives rise to functions $l_i : P\Omega \to \mathbb{F}_4$ obtained by sending a set A to $\sum_{x \in A \cap K_i} l(x)$. We now define the *6-tuple labeling map* $\mathcal{L}: P\Omega \to \mathbb{F}_4^6$ which sends A to the 6-tuple $(l_i(A))$. For example, the set

o	o				o
	o	o	o		o
			o		o
		o	o	o	

is labeled by $\left(01|\omega 0|\bar{\omega}\bar{\omega}\right)$. We call l_i the *i-th component of the labeling map.*

(5.13) Definition. A set A $\in P\Omega$ is *balanced* if the parity of $|A \cap K_i|$ is constant; this is called the *parity* of A. A set is *well-balanced* if balanced and the common parity is that of $|A \cap R_0|$. We call a set *even* or *odd*, if it is balanced and its parity is even or odd, respectively.

The condition that a set have intersections of common parity with two subsets S and T is a linear one since the map $i_{S,T}: A \mapsto \left(|A \cap S| + |A \cap T|\right) (mod\, 2) = \left(|A \cap (S+T)\,|\right) (mod\, 2)$ is linear. Thus, the families of balanced and well-balanced sets are linear subspaces. We leave as an exercise the verification that the linear functionals $i_{S,T}$ associated to $(S,T) = \left(K_1, K_j\right)$, $j = 2, \ldots, 6$ and (K_1, R_0) are linearly independent. The first five define a space $\mathcal{B}$ and all six define a space $\mathcal{W}$, whence $dim\, \mathcal{B} = 24 - 5 = 19$ and $dim\, \mathcal{W} = 24 - 6 = 18$. □

Note that the concept of hexacode balance (4.6.1) can be analyzed analogously by linear functionals.

(5.14) Lemma. *$\mathcal{L}$ maps $\mathcal{W}$ onto $\mathbb{F}_4^6$.*

Proof. Let (c_i) be a 6-tuple. The set S of 6 points with a single point in K_i at row c_i maps onto (c_i). If K is any column, $S + K$ and S have the same image and one of them is well-balanced.

(5.15) Definition. $\mathcal{G} := \mathcal{W} \cap \mathcal{L}^{-1}(\mathcal{H})$.

(5.16) Proposition. *$\mathcal{G}$ is a Golay code.*

Proof. The dimension is $18 - dim_{\mathbb{F}_2}\left(\mathbb{F}_4^6/\mathcal{H}\right) = 12$ and the length is 24, so by (5.9) all we need to do is prove that the value of d, the minimum weight, is at least 8. Suppose $A \in \mathcal{G}$ realizes the minimum weight. If A has odd parity, it has at least one point per column. The well-balanced requirement makes $A \cap R_0$ an odd set. If A has only 6 points, $\mathcal{L}(A)$ has oddly many zero coordinates, an impossibility for a hexacode word. Thus, A has weight at least $5 + 3 = 8$. If A has even parity, observe that $\mathcal{L}(A)$ has nonzero coordinate at i precisely when $A \cap K_i$ is a 2-set. So, $d \geq 2wt\,(\mathcal{L}(A))$. If $d < 8$, $wt\,(\mathcal{L}(A)) = 0$ and so A is a sum of evenly many columns (since $|A \cap R_0|$ is even), a contradiction. Therefore, $d \geq 8$. □

(5.17) Remark. We now have existence of $\mathcal{G}$. As an application of these methods, we show how to complete five points to an octad.

Take a 5-set, for instance one of $A_1, \ldots, A_5$, below:

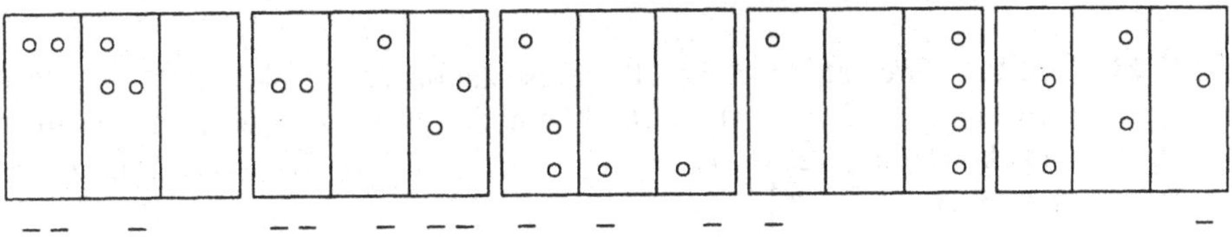

$$00\ 11\ 00 \quad 11\ 00\ \omega 1 \quad 01\ \overline{\omega}0\ \overline{\omega}0 \quad 00\ 00\ 00 \quad 0\omega\ 0\omega\ 01$$

Below each array, we have written the parity of the columns and the label of the set.

The easy case is that of equally many odd and even columns. The relevant octad is either odd or even. If odd, the three new points must go to the three even columns; if even, to the three odd columns. Use (4.6.3) to complete these three coordinates to a unique hexacode word and fill out the 5-set to the unique balanced 8-set with the proper label. A well balanced set arises in only one of the even or odd situation.

Suppose more than three columns have parity $p = +$ (for even) or $-$ (for odd). Then exactly 5 have that parity, say $\{K_i | i \in J\}$, for some 5-set $J \subseteq I = \{1, 2, \ldots, 6\}$. Let $\{t\} = I \backslash J$. Then one or three new points go to K_t. The zero or two which remain go to K_j, for some $j \in J$, and the label at that coordinate changes iff K_j gets two points. The distribution of the three new points thus depends on an error-correcting property of the hexacode; see (4.6.3).

Here are the completions of the above five sets. The correctness is easy to verify. One only checks that each set is well-balanced, is labeled by a hexacode word and has weight 8.

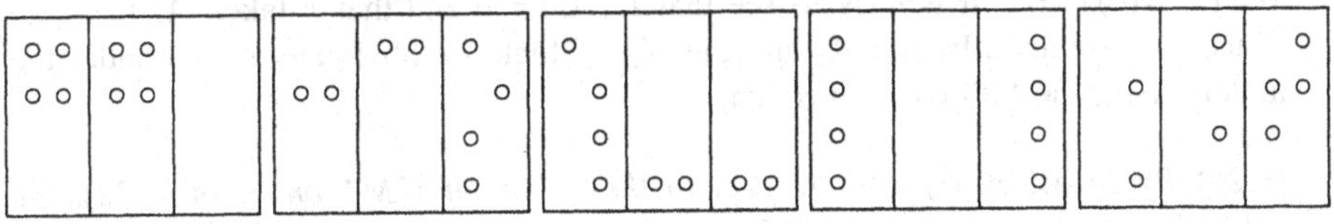

$$11\ 11\ 00 \quad 11\ 00\ 11 \quad 00\ \overline{\omega}\overline{\omega}\ \overline{\omega}\overline{\omega} \quad 00\ 00\ 00 \quad 0\omega\ 0\omega\ \overline{\omega}1$$

(5.18) Remark. It is worth observing that an even octad must be labeled by a hexacode word of weight 0 or 4.

(5.19) Notation. Write Ξ_u for the unordered sextet associated to Ξ. Let N be the subgroup of $Aut\,(\mathcal{G})$ which preserves Ξ_u ($Aut\,(\mathcal{G})$ is the group of the code $\mathcal{G}$ in the usual sense; it is a subgroup of Σ_Ω). There is a natural map $\pi_0 \colon N \to \Gamma := \Sigma_{\Xi_u}$; write L for its kernel.

(5.20) Notation. Let $V := \mathbb{F}_4^6$, and let $M = Mon\,(6,4)$, $M^* = Mon^*\,(6,4)$, $\mathcal{H}$, $Aut^*\,(\mathcal{H})$ have the meanings as in Section 3 and 4. We form the semidirect product $V{:}M^*$ and observe that it contains a natural subgroup $\mathcal{H}{:}\,Aut^*\,(\mathcal{H})$. Though V

is given as an abelian group under addition, we use multiplicative notation when referring to elements of V in VM^*.

(5.21) Notation. We let π be the HI-representation (1.21) of $V{:}M^*$ on a 24-set X associated to X_i, $i = 1,\dots,6$, where X_i is the 1-space in V spanned by the standard basis vector $e_i = (0,\dots,0,\underset{i}{1},0,\dots 0)$. As in (1.21), $X := \{(x,i) \mid x \in X_i, i = 1,\dots,6\}$.

The action is transitive and the X_i (identified with all (x,i), $x \in X_i$) form a system of imprimitivity. The stabilizer of $(0,i) \in X$ is the subgroup $Ker\,(p_i) : Stab_{M^*}(i)$ and it is an exercise that π is faithful.

(5.22) Notation. We define the set isomorphism $q{:}\,X \to \Omega$ by sending (X_i,i) to K_i in such a way that (ce_i,i) goes to the unique element of K_i labeled by $c \in \mathbb{F}_4$ by l; i.e., $q\,(ce_i) = x$ if $l\,(x) = c$. We thereby get from π an action π_1 of the group VM^* on Ω, defined by $q\,((ce_i,i))^{g^{\pi_1}} = q\left((ce_i,i)^{g^{\pi}}\right)$. Since π is faithful, so is π_1. Write $A{\cdot}v$ for the action of v on $A \in P\Omega$.

(5.23) Lemma. (i) *For $S \in P\Omega$, $v = (v_i) \in V$, we have $\mathcal{L}\,(S.v) = \mathcal{L}\,(S) + (c_i)$, where $c_i = |S \cap K_i| v_i$.* (ii) *If S is balanced, $\mathcal{L}\,(S.v) = \mathcal{L}\,(S) + p(S)v$, where $p(S)$ is the parity of S.* (iii) *The action of VM^* preserves the families of even and odd sets.* (iv) *For $v \in V$, the action of vm preserves the Golay code iff $v \in \mathcal{H}$.*

Proof. (i) Both sides are linear in S, so it suffices to check it for 1-sets. (ii) This follows trivially from (i). (iii) Trivial.

(iv) From (ii), "only if" is clear. We prove the "if" part. For each $c \in \mathcal{H}$, define a Golay set $A_c = K_1 + S_c$, where S_c is the 6-set which meets K_i in the 1-set $l^{-1}\,(c) \cap K_i$. It is easy to see that $\mathcal{L}\,(A_c) = c$ and that v takes A_c to A_{v+c}. Since $\mathcal{G}$ is spanned by the A_c and the K_{ij} (which are obviously stable under the action of v), the "if" part is proven. □

(5.24) Proposition. (i) *With respect to the action of VM^* on Ω in (5.22), the (global) stabilizer of $\mathcal{G}$ contains $\mathcal{H}.\,Aut^*\,(\mathcal{H})$;* (ii) *$\pi_0$ (5.19) is onto.*

Proof. (i) Use (5.23) and the fact that labelings stay in $\mathcal{H}$. (ii) Use (4.5.ii). □

(5.25) Proposition. (i) *N (the sextet stabilizer; see (5.19)) is isomorphic to $\left(\mathcal{H} : Aut^*\,(\mathcal{H})\right)^{\pi_1} \cong 2^6{:}[3{\cdot}\Sigma_6]$.*

(ii) *an element of N induces an even permutation on Ξ_u iff it is in $N' \cong 2^6 : \left[3^{\cdot}\,Alt_6\right]$.*

(iii) *$Stab_N\,(K_i) \cong 2^6{:}[3.\Sigma_5]$ induces Σ_4 on K_i and $g \in Stab_N\,(K_i)$ induces an even permutation on K_i iff $g \in N'$.*

Proof. (i) We use the notation of (5.19). Since $\left(Aut^*\,(\mathcal{H})\right)^{\pi}$ is Σ_{Ξ_u}, we need only show that $L = \langle \mathcal{H}, \mu \rangle$, where μ comes from the element "multiplication by ω" in $Aut^*\,(\mathcal{H})$, i.e.,

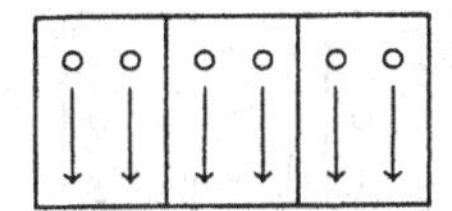

One containment is obvious, so we need only show that $L \leq \langle \mathcal{H}, \mu \rangle$. Let $g \in L$. Let z_i be the point labeled by 0 in K_i and consider $a_i = z_i^g, i = 1, \ldots, 6$. By (4.6.4), there is a unique hexacode word, w, which begins with $(l(a_1), l(a_2), l(a_3), \ldots)$. By replacing g with gw^{π_1}, we may assume g fixes z_1, z_2 and z_3. By considering the octad which contains $K_1 - \{z_1\}$, z_2 and z_3, we see that g fixes every point of Ω labeled with 0. Picture:

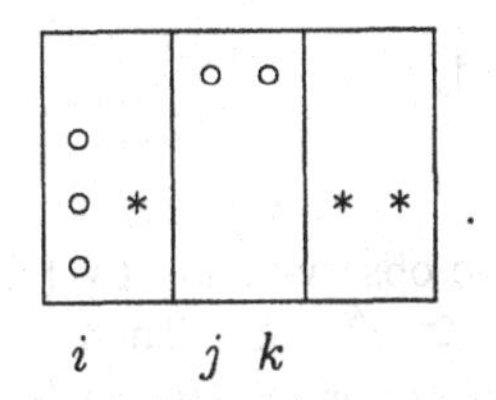

We claim that L_0, the subgroup of L fixing the points labeled with 0, acts faithfully on each K_i. Let $h \in L_0$ act trivially on some K_i. For $c \in \mathbb{F}_4^\times$, let x_c be the point labeled by c. The octad containing $K_i - \{x_c\}$ and a pair of points from K_j and K_k labeled with zero, where j and k form a coordinate block not containing i, is fixed by h, and since h preserves columns, is also fixed pointwise by h. Varying α, i and j, we get that h is trivial on Ω and so prove the claim. Picture:

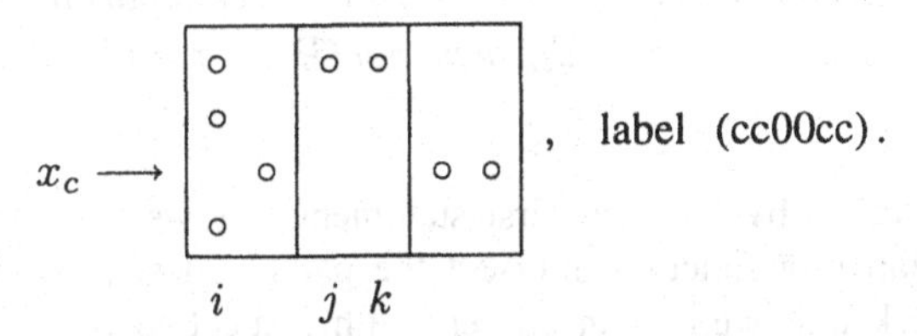

The claim proves that L_0 injects into Σ_3, via its action on any of the 3-sets $K_i^* := K_i - l^{-1}(0)$. Note thet μ induces a 3-cycle on each K_i^*. We may therefore assume that g induces a transposition on each K_i^*. If g fixes $x \in K_i$, $c := l(x) \neq 0$, expansion to an octad

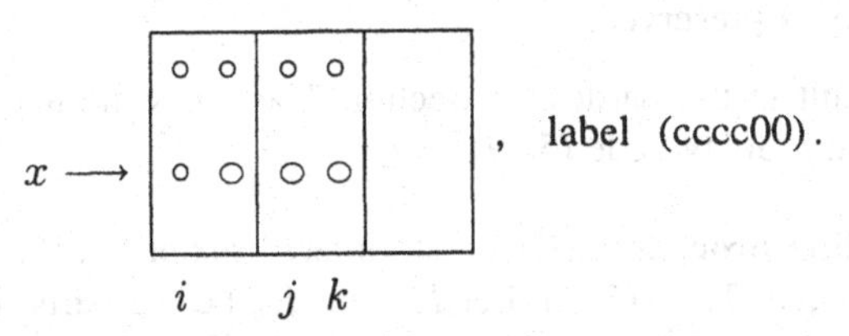

containing x and four points labeled with 0, all taken from columns in two coordinate blocks with one of these columns containing x, shows that g fixes every point of Ω labeled with c. Assuming the Proposition to be false, we replace g with a conjugate by a power of μ, we may assume that g looks like

o o	o o	o o
o o	o o	o o
↓ ↓	↓ ↓	↓ ↓

,

i.e., that g comes from the forbidden field automorphism of the extended monomial group! To get a contradiction, we just check that the octad

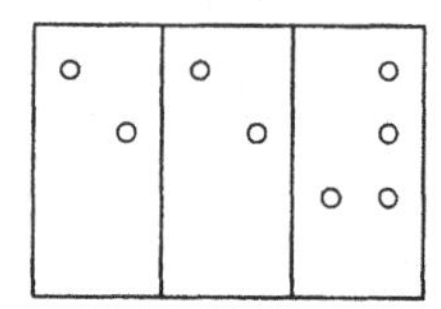

is sent outside $\mathcal{G}$ by g.

(ii) This is obvious once we observe that $\mathcal{H} = [\mathcal{H}, \mu]$.

(iii) Let $S := Stab_N(K_i) \cong 2^6{:}3.\Sigma_5$. Then, $S \cap N' \cong 2^6{:}3.\ Alt_5 \cong 2^6 : [3 \times Alt_5]$. Note that we get the action of Alt_4 on K_i by using elements of $\langle \mathcal{H}, \mu \rangle$ and we get an odd permutation by using the element α of (4.5). The group $S \cap N'$ induces Alt_4 on K_i. □

(5.26) Notation. Given an ordered partition $\mathcal{P}$ of Ω into 4-sets and a function l from Ω to $\mathbb{F}_4$ which restricts to a set isomorphism on each part of $\mathcal{P}$, we get a Golay code by the procedure of (5.15). Denote it by $\mathcal{G}(\mathcal{P}, l)$.

(5.27) Lemma. *Given pairs $(\mathcal{P}, l)$ and $(\mathcal{P}', l')$ of ordered partitions and labelings, there is a unique permutation $\sigma \in \Sigma_\Omega$ such that $(\mathcal{P}, l) \circ \sigma = (\mathcal{P}', l')$. Also, $\mathcal{G}(\mathcal{P}, l) \circ \sigma = \mathcal{G}(\mathcal{P}', l')$.*

Proof. This is fairly obvious. The first statement follows from the fact that the level sets of a labeling function intersect the parts in 1-sets. As for the second, we need to check that well-balanced sets with respect to one system go under σ to the same for the second system. Trivially, balanced sets go to balanced sets of the same parity. Let $S \in \mathcal{G}(\Xi, l)$ and $\mathcal{L}'$ the 6-tuple labeling map for $(\mathcal{P}', l')$. Then, $\mathcal{L}'(S \circ \sigma) = \mathcal{L}(S) \in \mathcal{H}$ and $(l^{-1}(0) \cap S) \circ \sigma = \{x\sigma \mid x \in S$ and $l(x) = 0\} = \{y \in S \circ \sigma \mid 0 = l(y\sigma^{-1}) = l'(y)\} = l'^{-1}(0) \cap S\sigma$, so the well-balanced property is preserved. □

The next result is the hardest in Section 5 and is central to the uniqueness proof of the binary Golay code (5.29).

(5.28) The labeling procedure. Given an ordered sextet, Ξ, we describe the 192 labelings l for which $\mathcal{G} = \mathcal{G}(\Xi, l)$. Let $K_1, \ldots, K_6$ be the parts, in order, and for $J = \{i, j, \ldots\} \subseteq \{1, \ldots, 6\}$, write K_J or $K_{ij\ldots}$ for $\sum_{i \in J} K_i$.

Let $\mathcal{X} := K_1 \times A_2 \times \mathbb{F}_4$, where $A_2 := \{(x, y) \in K_2 \times K_2 \mid x \neq y\}$ (the "antidiagonal"), and let $\mathcal{Y}$ be the set of labelings l (not yet proved to be nonempty) which satisfy $\mathcal{G} = \mathcal{G}(\Xi, l)$. Note that $|\mathcal{X}| = 192$.

For $p \in K_3$ and $(x, y) \in K_1 \times K_2$, let $\mathcal{O}(x, y, p)$ be the unique octad containing $(K_1 \setminus \{x\}) \cup \{y, p\}$. Since an intersection of two octads is an even set, $\mathcal{O}(x, y, p) \cap K_j$ is a 1-set for $j = 2, 3, 4, 5, 6$.

(5.28.1) If y and z are distinct elements of K_2, and p, $q \in K_3$, then $\mathcal{O}(x, y, p) \cap \mathcal{O}(x, z, q) \cap K_{3456}$ is a 1-set, for any p, $q \in K_3$.

(Reason: the intersection has at most 1 point since octads are determined by 5 points; it has exactly 1 point since $\{\mathcal{O}(x, y, p) \cap K_{3456} | p \in K_3\}$ partitions K_{3456}).

Throughout this labeling procedure, we fix an element $w \in K_3$. We assign to $l \in \mathcal{Y}$ the quadruple $(a, (b, c), \delta) \in \mathcal{X}$, where $l(a) = l(b) = 0$, $l(c) = 1$ and $\delta = l(w)$. This gives a map $\psi: \mathcal{Y} \to \mathcal{X}$, and we show that ψ is a bijection.

We shall prove first that ψ is an injection. Consider a labeling l which maps to $(a, (b, c), \delta)$ as above.

We argue that the values of l are determined on K_{3456}. This follows by noticing that $\mathcal{O}(a, c, w)$ must be labeled by the unique hexacode word of shape $(01|\delta\ldots)$ (4.6.4) in which every scalar in $\mathbb{F}_4$ occurs among its last four coordinates, then using (5.28.1) with $\mathcal{O}(a, c, w)$ and the octads $\mathcal{O}(a, b, q)$, $q \in K_3$, which must be labeled as $(00|\lambda\lambda|\lambda\lambda)$, to deduce l on K_{3456}.

By considering the set $\mathcal{N}$ of four octads different from K_{56} which contain $(l^{-1}(0) \cap K_5) \cup (K_6 \setminus l^{-1}(0))$, we see that the value of l is determined on K_{12} hence everywhere by the fact that such octads have label of the form $(\lambda\lambda\lambda\lambda 00)$. This proves injectivity of ψ.

Now to prove that ψ is onto. We are given $(a, (b, c), \delta) \in \mathcal{X}$ and we must find a labeling $l \in \mathcal{Y}$ such that $l^{\psi} = (a, (b, c), \delta)$. Recall that there is a fixed element $w \in K_3$ (see the definition of ψ).

Step 1. Let Ξ be the sextet containing the four set $T_1 := K_1 \setminus \{a\} \cup \{b\}$. Let $T_2 := K_{12} \setminus T_1$ and let $T_3, \ldots, T_6$ be the other parts of Ξ.

Step 2. Let $\mathcal{O}'' := \mathcal{O}(a, c, w)$. We expect the label to be the hexacode word $(01|\delta\delta'|\delta''\delta''')$. The last four scalars are distinct and are chosen to complete $(01|\delta\ldots)$ to a hexacode word (4.6.1). We define the label to be δ on w and $\delta', \delta'', \delta'''$ on the three new points (in the order $j = 3, 4, 5, 6$, according to membership in K_j). From (5.28.1), we see that $T_j \cap \mathcal{O}''$ is a singleton, for $j = 3, 4, 5, 6$, so we define l to be $\gamma \in \mathbb{F}_4$ at each point of T_j iff T_j contains the unique point of $\mathcal{O}''$ already given label γ. In this case, define $F_{\gamma} := T_j$.

Step 3. We now label K_{12}. For $s \in \mathbb{F}_4$, take a hexacode word of slope s (4.6) of the form $(0s|cd|ef)$; if $s \neq 0$, the scalars c, d, e, f are distinct. Let $U(cdef)$ be the 4-set consisting of the four points in $K_3 \cap F_c$, $K_4 \cap F_d$, $K_5 \cap F_e$ and $K_6 \cap F_f$. The other members of the sextet containing $U(cdef)$ are the three sets $U(dcfe), U(efcd), U(fedc)$, plus two further 4-sets $X_i(cdef)$, $i = 1, 2$, which have distributions $3 + 1$ and $1 + 3$ over K_1 and K_2. Index so that

$X_i := X_i(cdef)$ has 3 points in K_i, $i = 1,2$. Note that the 1-set $X_1(cdef) \cap K_2$ is unchanged if $cdef$ is replaced by $dcfe$, $efcd$ or $fedc$. For slope $t \neq s$, make similar definitions using the slope t hexacode word $(0t|c'd'|e'f')$; set $X_i' := X_i(c'd'e'f')$, $i = 1,2$.

We claim that $a \notin X_1$; for otherwise, $T_1 + F_0$ meets the octad $X_1 + U(cdef)$ in a 3-set if $s \neq 0$ and in a 7-set if $s = 0$, contradiction. Also, $a \notin X_1'$. The claim implies $X_1 = X_1'$ and $|U(klmn) \cap U(k'l'm'n')| = 1$, where $klmn$ is one of $cdef$, $dcfe$, $efcd$ or $fedc$ and $k'l'm'n'$ is one of $c'd'e'f'$, $d'c'f'e'$, $e'f'c'd'$ or $f'e'd'c'$. Note that $X_1(cdef) \cap K_2$ and $X_1(c'd'e'f') \cap K_2$ are distinct 1-sets; for otherwise, the intersection of the distinct octads $X_1(cdef) \cup U(cdef)$ and $X_1(c'd'e'f') \cup U(c'd'e'f')$ is a 5-set.

We now label the unique point of $X_1(cdef) \cap K_2$ with s (this is well-defined, by the previous sentence). As s varies, we label all of K_2. Note that b gets label 0. Similarly, we label the points of K_1 by using hexacode words of shape $(s0|cd|ef)$ and letting K_1 and $X_2(cdef)$ play the above roles of K_2, X_1.

Step 4. The last thing to check is that $(l(a), l(b), l(c)) = (0,0,1)$. Step 3 shows that a and b are labeled by 0. For c, it suffices to show that it is associated to a slope 1 situation as in Step 3; see Step 2 and note that the slope of $(01|\delta\delta'|\delta''\delta''')$ is 1.

We have produced a labeling l on Ω. Clearly, if $l \in \mathcal{Y}$, $l^{\psi} = ((a,(b,c),\delta))$. We let $\mathcal{L}$ be the associated 6-tuple labeling and define $\mathcal{G}(\Xi, l)$ as in (5.26). It remains to prove that $\mathcal{G} = \mathcal{G}(\Xi, l)$, which will imply that $l \in \mathcal{Y}$.

Notice that under $\mathcal{L}$, each octad K_{ij} gets label $(00|00|00)$ and that $T_1 + F_\gamma$ is labeled by the hexacode word $(00|\gamma\gamma|\gamma\gamma)$. The (8-dimensional) subspace H of $\mathcal{G}$ generated by these octads is contained in $\mathcal{G}(\Xi, l)$. Furthermore, for each slope s, we used an octad in $\mathcal{G}$ labeled by a hexacode word of shape $(0s|\ldots)$; the set $\mathcal{A}$ of all such octads lies in $\mathcal{G}(\Xi, l)$. Similarly, for each slope s, we used an octad in $\mathcal{G}$ labeled by $(s0|\ldots)$, and the set $\mathcal{B}$ of such lies in $\mathcal{G}(\Xi, l)$. The span of $\mathcal{A} \cup \mathcal{B}$ represents every coset of H in $\mathcal{G}$. Therefore, $\mathcal{G} \leq \mathcal{G}(\Xi, l)$. By dimensions, they are equal. This finishes verification of the labeling procedure.

(5.29) Proposition. *(Uniqueness of the Golay code) Given a Golay code $\mathcal{G}$ and a sextet Ξ, there exists a scalar labeling l so that $\mathcal{G} = \mathcal{G}(\Xi, l)$. In fact, there are* 192 *such labels. There is a bijection between* Aut $(\mathcal{G})$ *and pairs (Ξ, l), where l is a labeling such that $\mathcal{G} = \mathcal{G}(\Xi, l)$.*

Proof. (5.27) and (5.28). □

(5.30) Proposition. *Aut $(\mathcal{G})$ is transitive on sextets. The order of Aut $(\mathcal{G})$ is* $2^{10}3^3 5 \cdot 7 \cdot 11 \cdot 23 = 244{,}823{,}040$.

Proof. Use (5.29), then (5.25.i) and the fact that there are 1771 sextets; see after (5.10). □

(5.31) Proposition. (i) *The stabilizer of a four-set T induces Σ_T on it;* (ii) *$Aut\,(\mathcal{G})$ is 5-transitive on Ω.*

Proof. (i) A 4-set is part of a unique sextet, so without loss, we may take our four set to be K_1; then use (5.25.iii).

(ii) Let (x_i) and (y_i) be ordered sets of 5 points. We may take (x_i) to be as depicted below:

1 5		
2		
3		
4		

Let $T_1 = \{y_1, \ldots, y_4\}$ and expand to a sextet $\{T_1, \ldots, T_6\}$. Using transitivity on sextets, we may assume that this is the standard sextet. Using the action of the sextet group (see (5.25)) as Σ_6 on tetrads, we may assume that $T_1 = K_1$ and $y_5 \in K_2$. From (i), we may assume $x_i = y_i$, for $i = 1, 2, 3, 4$. Finally, the permutation representation of codewords on Ω (see (5.21)) allows us to move y_5 to x_5 by using a scalar multiple of a codeword $(0101\omega\bar{\omega})$. □

(5.32) Definition. Let $\mathcal{G}$ be a Golay code. Define the *Mathieu group* M_{24} to be $Aut\,(\mathcal{G})$. Due to the uniqueness properties of $\mathcal{G}$ (5.29), M_{24} is well-defined, up to conjugacy, in Σ_{24}.

(5.33) Proposition. *M_{24} is a simple and 5-fold transitive permutation group on 24 points.*

Proof. The group order, (1.18) and Sylow's theorem give 23:11 as the structure of the Sylow 23-normalizer of $G = M_{24}$. For a p-subgroup P of M_{24}, $23 \nmid |Aut\,(P)|$ by (1.17). Now consider a minimal normal subgroup K of G, $S \in Syl_{23}\,(G)$ and $S \cap K$. Since G is multiply transitive it is primitive (1.2), whence K is transitive (1.3.i); therefore, 24 divides $|K|$.

Suppose that $S \leq K$, the Frattini argument tells us that $N_G\,(S)\,K = G$. If $S = N_K\,(S)$, Burnside's theorem tells us that $K = O_{23'}\,(K)\,S$; but since K is minimal normal, $K = O_{23'}\,(S)$ and so S is not contained in K, a contradiction. So, $S < N_K\,(S)$, whence $N_K\,(S) = N_G\,(S)$ since the latter group has order 23.11. The Frattini argument now says that $K = G$. So, G is simple.

Suppose finally that $S \cap K = 1$. Then, the Schur-Zassenhaus theorem (1.9) and the fact that 3 divides $|K|$ and $|M_{24}| = 3^3$ implies that S centralizes a Sylow 3-subgroup of K ((1.17), $p = 3$, $d = 3$). But this contradicts the structure of $N_G\,(S)$, referred to above. □

(5.34) Remark. M_{24} is not 6-transitive, or else by (1.3), 19 would divide $|M_{24}|$, against (5.31). Actually, one can prove that $Aut\,(\mathcal{G})$ is not 6-transitive by using (1.4) (without (1.8) and knowing $|Aut(\mathcal{G})|$).

(5.35) Standard Basis of $\mathcal{G}$:

(5.36) A few useful sextet labelings. In each case, to verify that the labelings give our standard Golay code $\mathcal{G}$, we check that the standard basis for $\mathcal{G}$ is in the indicated $\mathcal{G}(\Xi, l)$. For future reference, the standard labeling is $\mathcal{SL}_0$ and the ones which follow are $\mathcal{SL}_m$, $m = 1, 2, \ldots$. The order of the parts of the sextet is indicated by the small integer, placed centrally if possible, or by doubling the symbols in the second of the two parts within a given K_{12}, K_{34} or K_{56}. After the validity of $\mathcal{SL}_4$ is verified, the validity of $\mathcal{SL}_k$ follows, for $1 \leq k \leq 6$; see (6.8.4).

$\mathcal{SL}_1$					
0	$\overline{\omega}$	0	$\overline{\omega}$	0	$\overline{\omega}$
11	$\omega\omega$	11	$\omega\omega$	11	$\omega\omega$
$\overline{\omega}\overline{\omega}$	00	$\overline{\omega}\overline{\omega}$	00	$\overline{\omega}\overline{\omega}$	00
ω	1	ω	1	ω	1

$\mathcal{SL}_2$					
0	ω	0	ω	0	ω
$\omega\omega$	00	$\omega\omega$	00	$\omega\omega$	00
1	$\overline{\omega}$	1	$\overline{\omega}$	1	$\overline{\omega}$
$\overline{\omega}\overline{\omega}$	11	$\overline{\omega}\overline{\omega}$	11	$\overline{\omega}\overline{\omega}$	11

$\mathcal{SL}_3$					
0	$\overline{\omega}\overline{\omega}$	0	$\overline{\omega}\overline{\omega}$	0	$\overline{\omega}\overline{\omega}$
00	$\overline{\omega}$	00	ω	00	$\overline{\omega}$
11	ω	1	ω	1	ω
1	$\omega\omega$	1	$\omega\omega$	1	$\omega\omega$

$\mathcal{SL}_4$					
0	1	0	1	0	1
1		3		5	
$\overline{\omega}$	ω	$\overline{\omega}$	ω	$\overline{\omega}$	ω
ω	$\overline{\omega}$	ω	$\overline{\omega}$	ω	$\overline{\omega}$
2		4		6	
1	0	1	0	1	0

$\mathcal{SL}_5$					
0	11	0	11	0	11
ω	$\overline{\omega}\overline{\omega}$	ω	$\overline{\omega}\overline{\omega}$	ω	$\overline{\omega}\overline{\omega}$
00	1	00	1	00	1
$\omega\omega$	$\overline{\omega}$	$\omega\omega$	$\overline{\omega}$	$\omega\omega$	$\overline{\omega}$

$\mathcal{SL}_6$					
0	$\omega\omega$	0	$\omega\omega$	0	$\omega\omega$
$\overline{\omega}\overline{\omega}$	1	$\overline{\omega}\overline{\omega}$	1	$\overline{\omega}\overline{\omega}$	1
$\overline{\omega}$	11	$\overline{\omega}$	11	$\overline{\omega}$	11
00	ω	00	ω	00	ω

(5.37) Criterion for a permutation to be in M_{24}. If $g \in \Sigma_{24}$, $g \in M_{24}$ iff $\mathcal{G}g = \mathcal{G}$. Since $\mathcal{G}$ is self orthogonal, this is equivalent to the following: $|Ag \cap B|$ is even, for all A, B in a basis for $\mathcal{G}$; see (5.36). By taking a basis which contains Ω, it suffices to let A and B range over the other 11 elements of the basis. This is straightforward to check by hand. A computer program is useful here.

Another method is to transform the labeling by g, then verify that members of the standard basis satisfy the defining conditions of the new Golay code.

(5.38) Some useful permutations. We shall refer to these as UP0, UP1,... . Membership in M_{24} may be verified with (5.38), although in a few cases there is a shorter method discussed in the proofs of Section 6. We do not list here many elements of M_{24} we use which are described in an obvious way using the standard sextet labeling.

	Diagram	**Comments**
UP0	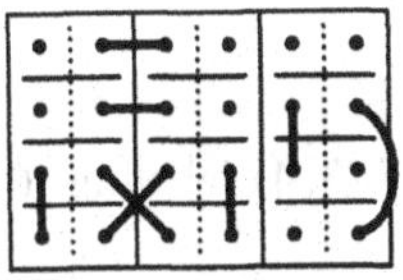	corresponds to our element of $Aut^*(\mathcal{H})$ inducing (23) on the parts of the sextet; (4.5)
UP1	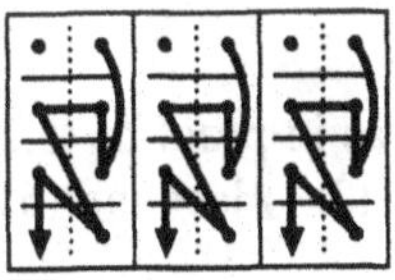	natural element of order 7 in $G(\mathcal{O}) \cap G(\mathcal{T})$ acting as $x \mapsto x+1$; (6.8)
UP2	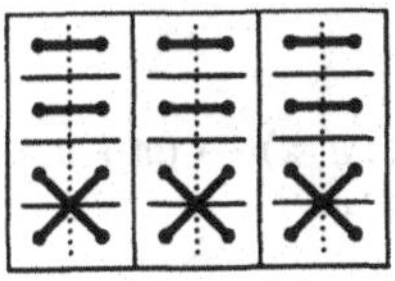	lives in group of the sextet, octad, trio and in $L(2,7)$, $PGL(2,11)$, $L(2,23)$; (6.8), (6.14), (6.19); notated τ in (6.20)
UP3	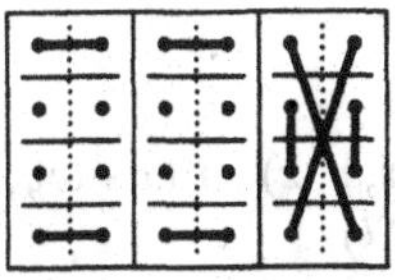	acts as $(\bar{\omega}0|\bar{\omega}|1\omega)$ w.r.t. $\mathcal{SL}_1$ (6.36)
UP4	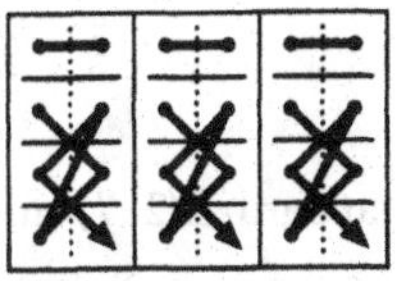	product of μ (= UP9) with $r_{(12)(34)}$; in sextet group (5.25)

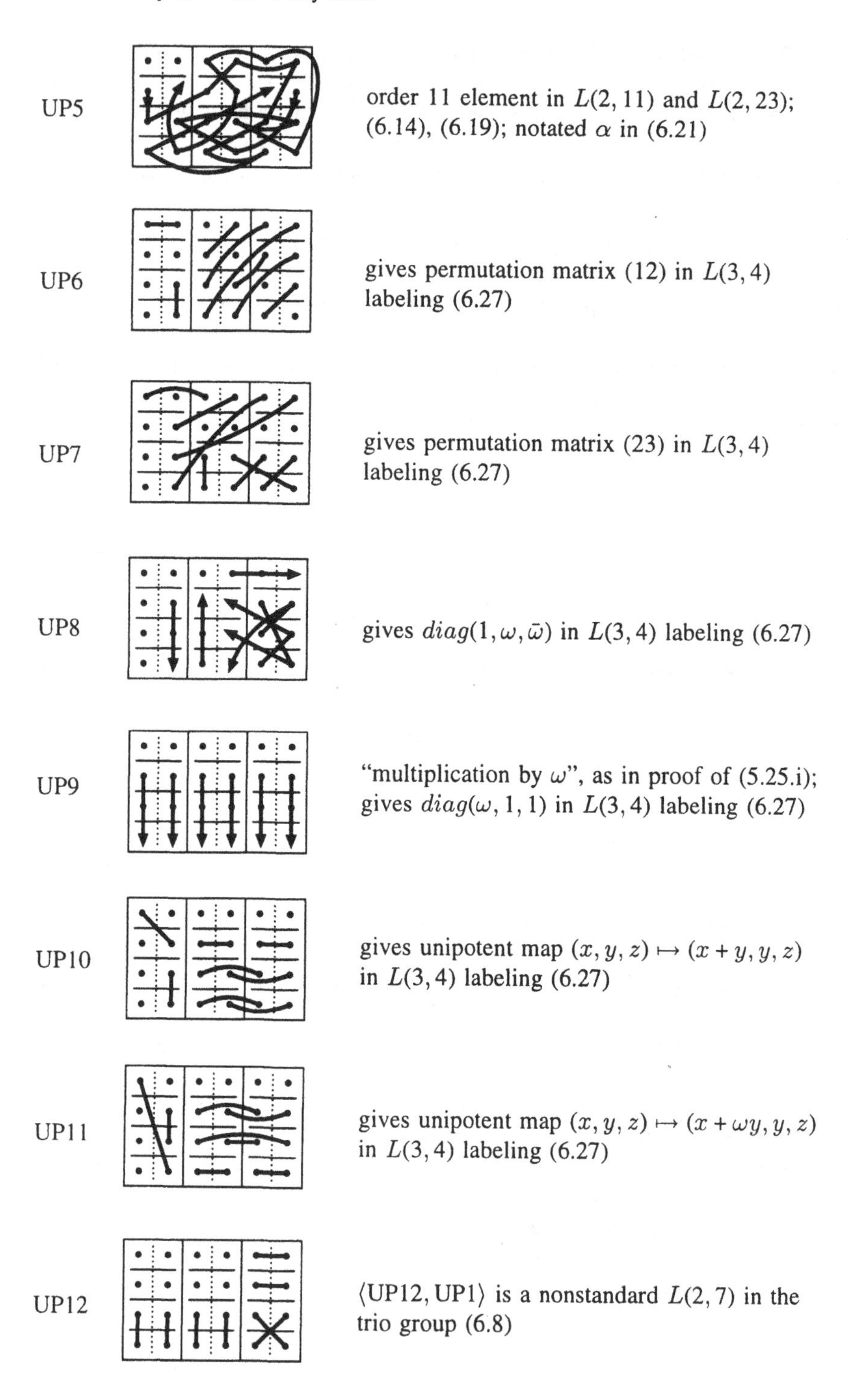
UP5
order 11 element in $L(2,11)$ and $L(2,23)$; (6.14), (6.19); notated α in (6.21)
UP6
gives permutation matrix (12) in $L(3,4)$ labeling (6.27)
UP7
gives permutation matrix (23) in $L(3,4)$ labeling (6.27)
UP8
gives $diag(1,\omega,\bar{\omega})$ in $L(3,4)$ labeling (6.27)
UP9
"multiplication by ω", as in proof of (5.25.i); gives $diag(\omega,1,1)$ in $L(3,4)$ labeling (6.27)
UP10
gives unipotent map $(x,y,z) \mapsto (x+y,y,z)$ in $L(3,4)$ labeling (6.27)
UP11
gives unipotent map $(x,y,z) \mapsto (x+\omega y,y,z)$ in $L(3,4)$ labeling (6.27)
UP12
$\langle$UP12, UP1$\rangle$ is a nonstandard $L(2,7)$ in the trio group (6.8)

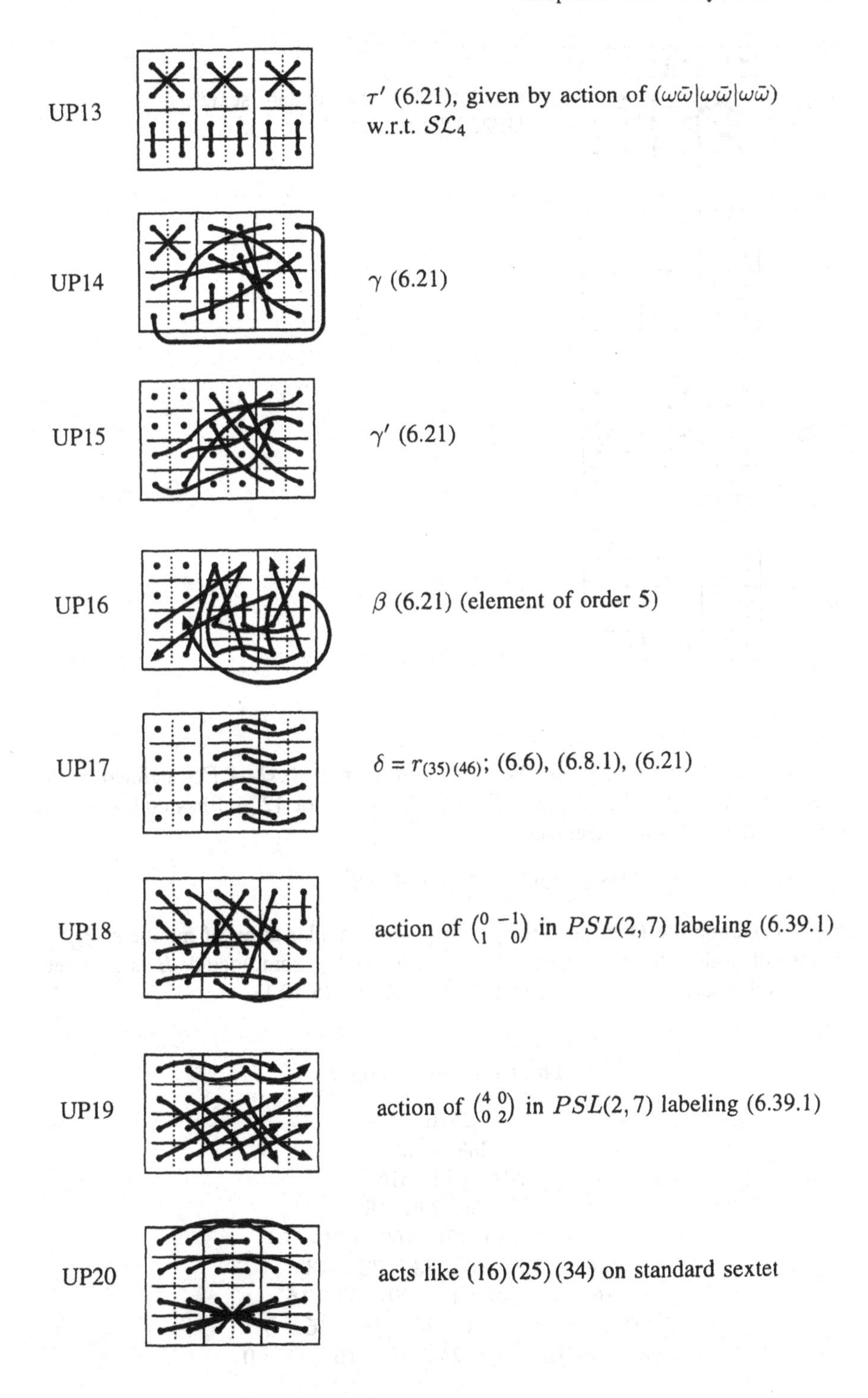

UP13 τ' (6.21), given by action of $(\omega\bar{\omega}|\omega\bar{\omega}|\omega\bar{\omega})$ w.r.t. $\mathcal{SL}_4$

UP14 γ (6.21)

UP15 γ' (6.21)

UP16 β (6.21) (element of order 5)

UP17 $\delta = r_{(35)(46)}$; (6.6), (6.8.1), (6.21)

UP18 action of $\binom{0\ -1}{1\ \ \ 0}$ in $PSL(2,7)$ labeling (6.39.1)

UP19 action of $\binom{4\ 0}{0\ 2}$ in $PSL(2,7)$ labeling (6.39.1)

UP20 acts like (16)(25)(34) on standard sextet

UP21 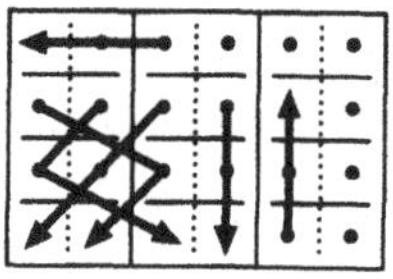 acts like (321) on standard sextet; UP20.UP18.UP20

UP22

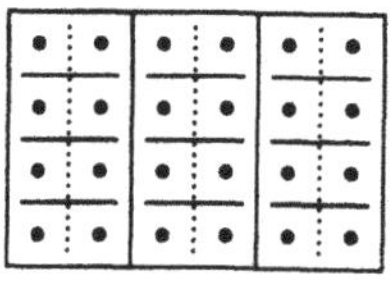

UP23

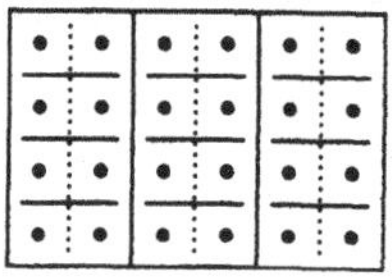

UP24

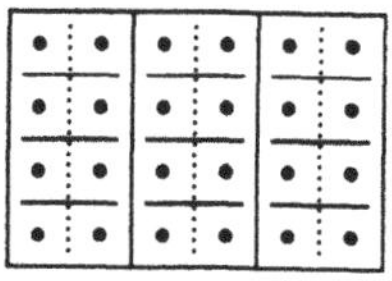

(5.40) Exercises

(5.40.1) Let S be a Golay set. We have a linear map $i_S: \mathcal{G} \to PS$, defined by A $\mapsto A \cap S$. Show that $Im\,(i_S)$ is $PE\,(S)$ if $|S| = 0, 8$ or 12 and has codimension 4 in $PE\,(S)$ if S is a sixteen-set.

Hint: analyze $Ker\,(i_S)$, using the Leech triangle.

(5.40.2) Define the Dodecad Triangle to be a triangular array of numbers Q_{ij} = number of dodecads which meet B in A (where $B \subseteq$ some octad, B is an i-set and A is a j-set). Hint: Use the Octad Triangle and (5.40.1).

The Dodecad Triangle

2576
1288 1288
616 672 616
280 336 336 280
120 160 176 160 120
48 72 88 88 72 48
16 32 40 48 40 32 16
0 16 16 24 24 16 16 0
0 0 16 0 24 0 16 0 0

(5.40.3) Let $\mu = UP9$ (5.38). Let $\mathcal{G}'$ be the set of Golay codewords supported on E, the 18-set which is the union of the 6 nontrivial orbits of μ. Define an $\mathbb{F}_2$-linear map φ of $P\Omega$ to $\mathbb{F}_4^6$ by sending the unique element of $R_c \cap K_i$ to ce_i. Prove that $\mathcal{G}^\varphi = \mathcal{G}'^\varphi = [\mathcal{G}, \mu]^\varphi = \mathcal{H}$ and that φ is a map of N-modules (see (5.25)), where N acts on $\mathbb{F}_4^6$ as $Aut^* \mathcal{H}$. Show that $\varphi|_{\mathcal{G}'}$ is an isomorphism. This is somewhat reminiscent of code thickening (3.19).

(5.41) Exercise. Prove that $\mathcal{G}$ is a uniserial module for M_{24}, with ascending composition factors of dimensions 1 and 11. Determine the module structure for subgroups of the forms 23, 23:11, 11, 11:5. Hints: (1) If H is a cyclic group of odd order, $\mathbb{F}_2 H$ is a direct sum of fields. (2) To prove that the trivial submodule $\langle \Omega \rangle$ is not complemented by an M_{24}-module, use an element t which interchanges a dodecad and its complement to show that $\Omega \in [\mathcal{G}, t]$.

Chapter 6. Subgroups of M_{24}

We discuss some important subgroups of M_{24}, including the maximal subgroups. An important early paper on this matter is that of John Todd [To], who gives character tables for many important subgroups and lists all 759 octads! (though a few errors appeared on that list). The thesis of Chang Choi [Ch] claimed a classification of maximal subgroups of M_{24}, but missed the (then unknown) transitive and imprimitive $PSL(2,7)$-subgroup. The later paper of Robert Curtis [Cu] shows that there are nine classes of maximal subgroups.

We shall describe these nine classes of maximal subgroups. In several of those cases, we give a lot of detail, especially when the treatment can be closely connected with the ideas and notations of Sections 4 and 5. Here, we mention that "the Mathieu groups" usually means the five simple groups M_{11}, M_{12}, M_{22}, M_{23}, M_{24}; references to $M_{10} \cong Alt_6 \cdot 2$, $M_9 \cong 3^2{:}Quat_8$ and $M_{21} \cong PSL(3,4)$ do occur in the literature and we discuss these groups in this section.

(6.1) Notation. If H is a group operating on the finite set X, we say its *orbit decomposition* is $a+b+\ldots$ if the action of H on X has orbits of lengths $a, b, \ldots$.

(6.2) Notation. In Section 6, G denotes the group M_{24}. Let $\mathcal{T}$ be a *trio* of octads, i.e., three pairwise disjoint octads; they partition Ω. Now let X be one of: an octad, a trio or a sextet. The group $G(X)$ is the stabilizer of X in G, $R(X) = O_2(G(X))$ and $L(X)$ denotes a particular complement to $R(X)$ in $G(X)$ (it is unique, up to conjugacy). These groups are indicated in Table 6.1. Unless stated explicitly to the contrary, Ξ is the fixed sextet used throughout Section 5 and we freely refer to its associated picture and labeling. For $\mathcal{O}$, we usually take K_{12} and for $\mathcal{T}$, we usually take $\{K_{12}, K_{34}, K_{56}\}$.

(6.3) Maximal Subgroups of M_{24} (see [Cu])

Transitive Subgroups	Intransitive Subgroups (orbit lengths)	
$M_{12}{:}2$ (dodecad pair group)	M_{23}	$(1+23)$;
$PSL(2,23)$;	$M_{22}{:}2$	$(2+22)$;
$G(\Xi)$ (sextet group)	$M_{21}{:}\Sigma_3 \cong P\Gamma L(3,4)$	$(3+21)$;
$R(\Xi) \cong 2^6$,		
$L(\Xi) := Aut^*(\mathcal{H})^{\pi_1} \cong 3{\cdot}\Sigma_6$;		
$PSL(2,7)$;	$G(\mathcal{O})$ (octad group)	$(8+16)$
$G(\mathcal{T})$ (trio group)	$R(\mathcal{O}) \cong 2^4$, $L(\mathcal{O}) \cong GL(4,2)$,	
$R(\mathcal{T}) \cong 2^6$,	$G(\mathcal{O}) \cong AGL(4,2)$.	
$L(\mathcal{T}) \cong \Sigma_3 \times PSL(2,7)$.		

The only primitive subgroup in the table is $PSL(2,23)$. The transitive groups listed leave invariant systems of imprimitivity with block sizes 12, 4, 8 and 3. There are interesting systems with block sizes 2 and 6, but the groups leaving them invariant are not maximal. It is easy to show that any intransitive group is on the above list; for if the subgroup H has an orbit S, either S is a Golay set or S is in a coset of $\mathcal{G}$ of weight at most 4; in the latter case, H fixes a k-set, for some $k \leq 3$ or a sextet.

(6.4) Labelings with respect to certain proper subgroups. (These will be discussed as the section unfolds).

The sextet labeling

0	0	0	0	0	0
1	1	1	1	1	1
ω	ω	ω	ω	ω	ω
$\overline{\omega}$	$\overline{\omega}$	$\overline{\omega}$	$\overline{\omega}$	$\overline{\omega}$	$\overline{\omega}$

The octad labeling

*	*	
*	*	$\mathbb{F}_2^4$
*	*	
*	*	

The trio labeling

∞	0	∞	0	∞	0
3	2	3	2	3	2
5	1	5	1	5	1
6	4	6	4	6	4

The $PGL(2,11)$ labeling

∞	∞	3	3	4	4
0	0	2	2	X	X
1	8	9	6	5	7
8	1	6	9	7	5

R	L	R	L	R	L
L	R	L	R	L	R
L	R	L	R	L	R
L	R	L	R	L	R

The $PSL(2,23)$ labeling

∞	0	22	1	11	2
3	15	12	21	13	7
6	5	18	20	4	10
9	19	8	14	16	17

The $PSL(3,4)$ labeling

∞ 0	00 10	$\omega 0$ $\overline{\omega}0$
I 1	01 11	$\omega 1$ $\overline{\omega}1$
II ω	0ω 1ω	$\omega\omega$ $\overline{\omega}\omega$
III $\overline{\omega}$	$0\overline{\omega}$ $1\overline{\omega}$	$\omega\overline{\omega}$ $\overline{\omega}\,\overline{\omega}$

mnemonic:

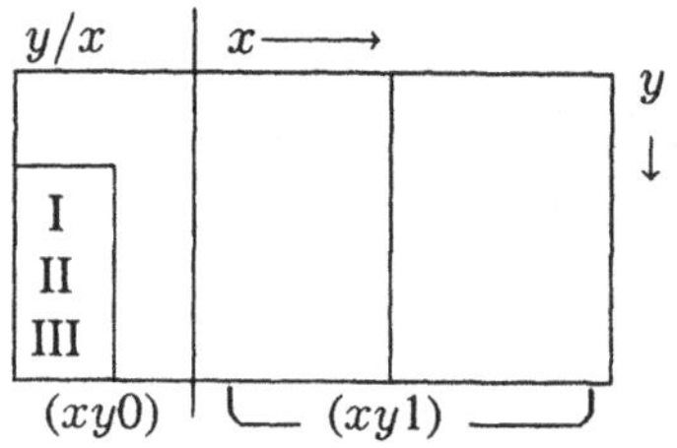

y/x | $x \longrightarrow$

y $\downarrow$

I
II
III

$(xy0)$ | $(xy1)$

The $L_2(7)$-labeling

10 51	30 62	20 43
11 01	52 32	63 23
31 61	22 12	53 03
41 21	42 02	13 23

$= \left[\mathbb{F}_7^2 \backslash \{0\}\right] / \{\pm 1\}.$

(6.5) Exercises

(6.5.1) Determine the structure of a Sylow 5-centralizer and 5-normalizer in $G = M_{24}$. Hints: (i) Let $x \in G$ have order 5. By taking such an x in a sextet stabilizer, H, see that its cycle structure must be $1^4 5^4$; (ii) Get $C_H(x) \cong 5 \times Alt_4$ and $N_H(\langle x \rangle) \cong [5 \times Alt_4]{:}4 \cong [Dih_{10} \times Alt_4]\cdot 2 \cong [Frob(20) \times \Sigma_4]\frac{1}{2}$ (see (2.1)); (iii) if T denotes the 4-set of fixed points of x, show that $N_G(\langle x \rangle)$ stablilizes T, hence stabilizes the sextet containing T and so $N_G(\langle x \rangle) = N_H(\langle x \rangle)$.

(6.5.2) Show that a sextet stabilizer is a maximal subgroup. Hint: if H is a sextet stabilizer, its index is $1771 = 7{\cdot}11{\cdot}23$; Sylow theory and (6.5.1) imply that any subgroup between H and G has index $\equiv 1 (mod 5)$.

(6.6) Notation. $r_g = g\pi_1$, for $g \in V{:}Mon^*(6,4)$; see (5.23). In case g is in V or is a permutation matrix, we may write the coordinates or cycle structure, respectively, in place of g.

(6.7) Lemma (Commuting criterion). *Let X be a set partitioned into subsets $\{X_i \mid i \in I\}$ and let s_1 and s_2 be involutions in Σ_X. Suppose each X_i is invariant under $\langle s_1, s_2 \rangle$, has cardinality* 1, 2 *or* 4, *and that for each i and j, s_j is trivial or fixed point free on X_i. Then s_1 and s_2 commute.*

Proof. This boils down to a property of Σ_4. □

(6.8) Theorem (Structure of an octad stabilizer). *G has one orbit on octads. Let $H := G(\mathcal{O})$ be the stabilizer in G of an octad, $\mathcal{O}$. Then, $H \cong AGL(4,2)$, the semidirect product of an elementary abelian group of order 16 by its automorphism group $GL(4,2)$. It induces as the alternating group $Alt_{\mathcal{O}}$ on the 8 points of $\mathcal{O}$, the kernel of the action is $R(\mathcal{O}) := O_2(H)$, and H acts faithfully on $\mathcal{O} + \Omega$ with $O_2(H)$ as a regular normal subgroup. Consequently, $Alt_8 \cong GL(4,2)$ (one of the "exceptional isomorphisms" among finite simple groups).*

Proof. Since G is 5-transitive on Ω, the Steiner system axioms imply transitivity on octads. We take $\mathcal{O} = K_1 + K_2$ and adopt the notation $r_\alpha, \alpha \in \mathbb{F}_4$, for the permutation $r_{(00|\alpha\alpha|\alpha\alpha)}$ and r_{ijkl} for r_σ, where $\sigma = (ij)(kl) \in S$ (see (6.6)).

We define R to be the group generated by the permutations r_α and r_σ for $\sigma \in \langle r_{3456}, r_{3546} \rangle$, a four-subgroup in S; see (4.1). We give a picture of r_1, r_ω, r_{3456} and r_{3546} below:

(6.8.1) 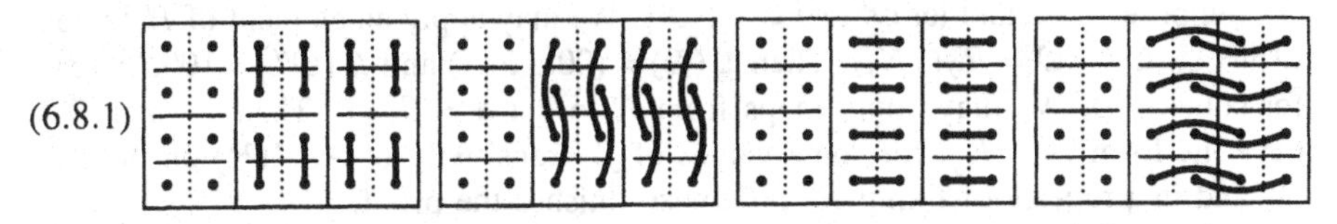

Using the technique of (6.7), it is readily verified that R is an elementary abelian group of order 16, having the four involutions above as a basis, and that R acts regularly on $\Omega + \mathcal{O}$. Since it is in the sextet group $G(\Xi)$, it is in G, hence in H.

Write π for the permutation representation of H on $\mathcal{O}$. Certainly, R is in the kernel of π. Since G is 5-transitive, H is 5-transitive on $\mathcal{O}$, so $N := |H^\pi|$ is divisible by 8.7.6.5.4. (Note that (1.4) implies that $N \cong Alt_8$ or Σ_8). We claim that N is divisible by 3^2. This is obvious from comparing how the permutation μ

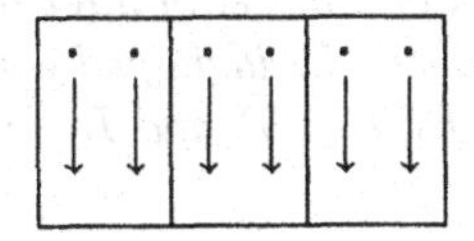

operates on the two octads

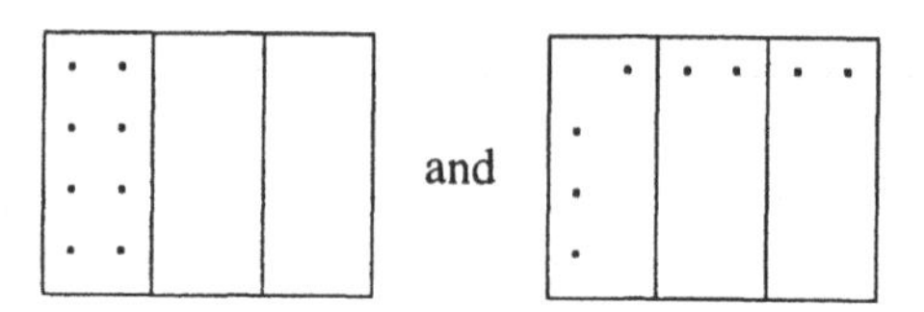

Thus, $N = 8!$ or $8!/2$. Since $|H| = 16.(8!/2)$, $R = Ker\,\pi$ and $H^{\pi} = Alt_{\mathcal{O}}$.

It remains to prove the isomorphism of H with the affine group $AGL(4,2) \cong 2^4{:}\,GL(4,2)$. As remarked above, R operates regularly on Ω, so a point stabilizer complements R in H. Finally, since R is not in the center of H (for instance, μ commutes with none of the given generators of R), the action of H on R is nontrivial. This gives a nontrivial map of H/R into $Aut(R)$. Since $H/R \cong Alt_8$ is simple, the map is monic. Since $Aut(R)$ has order $8!/2$, the map is an isomorphism and so $Alt_8 \cong GL(4,2)$. □

(6.9) Corollary. *G is transitive on the sets of pairs of octads with a given cardinality of intersection. In particular, G is transitive on trios and on the set of ordered trios.*

Proof. It suffices to show that $H = G(\mathcal{O})$ is transitive on the sets $X_n := \{\mathcal{O}' \in \mathcal{S} \mid \mathcal{O}' \cap \mathcal{O}$ is an n-set$\}$, for $n = 0$, 2 and 4.

Suppose $n = 0$. The Leech triangle tells us that $|X_n| = 30$. Identify $\Omega \backslash \mathcal{O}$ with $\mathbb{F}_2^4$ by choosing an "origin," say the point z at the upper left, then using the action of R. Let R_1 be the eights group generated by the first three maps in (6.8.1). Then $z^{R_1} = K_{34}$, which means that the octad $K_{34} \in X_0$ is a 3-dimensional affine subspace. The action of H preserves octads and is transitive on the set of 30 rank 3 affine subspaces, hence transitive on X_0.

Suppose $n = 2$ and let $\mathcal{O}'$ and $\mathcal{O}'' \in X_2$. We may apply an element of H to get $\mathcal{O}' \cap \mathcal{O} = \mathcal{O}'' \cap \mathcal{O} = R_0 \cap K_{12}$. Then $\mathcal{L}(\mathcal{O}) = (00\alpha\alpha\alpha\alpha)$ and $\mathcal{L}(\mathcal{O}'') = (00\beta\beta\beta\beta)$, for some α, β. We may apply maps from R to get $\alpha = \beta = 0$. Then $\mathcal{O}'$ and $\mathcal{O}''$ have the form $R_0 + K_j$, for some values of j. Transitivity of $Aut^*(\mathcal{H})$ on the six K_i and its property of fixing R_0 pointwise finishes the proof.

Finally, suppose $n = 4$. We may assume that all three octads contain K_1 and furthermore that $\mathcal{O}' = K_{13}$ and $\mathcal{O}'' = K_{14}$. The 2-transitivity of $G(\Xi)$ on the six K_i finishes the proof. □

(6.10) Theorem (Structure of a trio stabilizer)**.** *Let $H = G(\mathcal{T})$. Then, H is the semidirect product of $R := O_2(H) \cong 2^6$ by the group $L = L_1 \times L_2$ acting faithfully, where $L_1 \cong GL(3,2)$, $L_2 \cong \Sigma_3$. The subgroup of H fixing each octad of the trio is $O_2(H)L_1$ and L_2 induces Σ_3 on this set of three octads. The group L_1 fixes no points of Ω. In fact, each octad may be identified with the projective line over $\mathbb{F}_7$ in such a way that L_1 acts as $PSL(2,7)$ and L_2 faithfully permutes the octads, preserving the identification.*

Proof. We take the trio of octads to be $K_{i,i+1}$, for $i = 1, 3, 5$, in the usual sextet labeling convention. Call them $\mathcal{O}_j$, $j = 1, 2, 3$, respectively. Define $G(i) := G(\mathcal{O}_i)$ and $R(i) := R(\mathcal{O}_i)$, for $i = 1, 2, 3$.

First, we define the subgroup R of H as the group generated by permutations a_i, b_i, c_i, for $i = 1, 2, 3$. These permutations are defined as

$$\begin{aligned} a_1 &= r_{(00|11|11)}, \quad a_2 = r_{(11|00|11)}, \quad a_3 = r_{(11|11|00)}; \\ b_1 &= r_{(00|\omega\omega|\omega\omega)}, \quad b_2 = r_{(\omega\omega|00|\omega\omega)}, \quad b_3 = r_{(\omega\omega|\omega\omega|00)}; \\ c_1 &= r_{(34)(56)}, \quad c_2 = r_{(12)(56)}, \quad c_3 = r_{(12)(34)}. \end{aligned} \tag{6.10.1}$$

They may be given by the following respective diagrams:

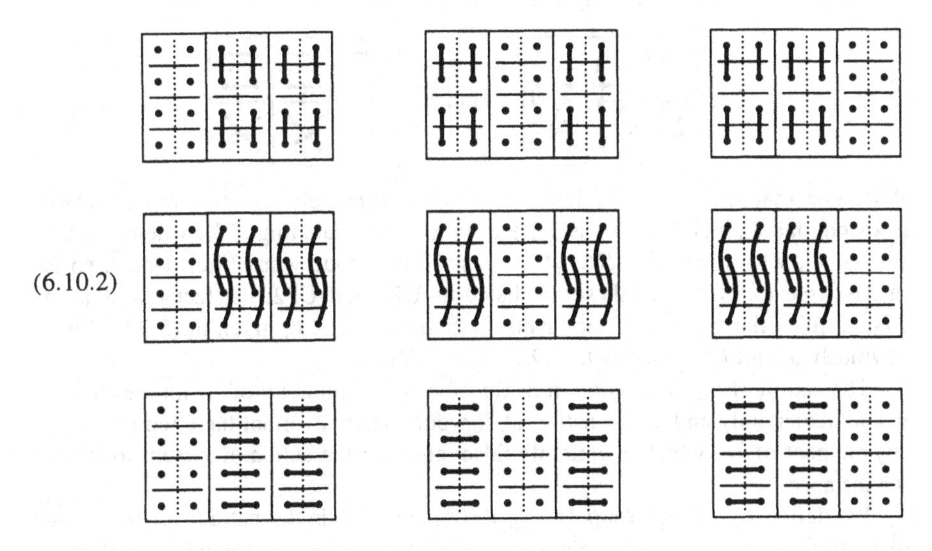

(6.10.2)

The technique of (6.7) may be used to show that any two of these elements commute. Thus, R is an elementary abelian 2-group of order at most 2^9. Notice that $a_1a_2a_3 = b_1b_2b_3 = c_1c_2c_3 = 1$. These relations imply at once that $|R|$ divides 2^6.

Define $R_i := \langle a_i, b_i, c_i \rangle$, for $i = 1, 2, 3$. This is a subgroup of $R(i)$ whose structure is described after (6.8.1); R_i has rank 3 and index 2 in $R(i)$. Let $\{i, j\}$ be a 2-set in $\{1, 2, 3\}$. Since $R(i)$ fixes $\mathcal{O}_i$ pointwise and acts regularly on $\mathcal{O}_i + \Omega$, $R_i \cap R_j = 1$, so that $R = R_iR_j = R_i \times R_j$ is elementary abelian of order 64. This is so for all such $\{i, j\}$. Therefore, R is normal in H.

We label the blocks with $\mathbb{P}^1(7)$ as follows:

(6.10.3)

∞	0	∞	0	∞	0
3	2	3	2	3	2
5	1	5	1	5	1
6	4	6	4	6	4

.

We define L_1 as the group acting by the same linear fractional transformations on each block simultaneously; recall that the *left* action of an element of $GL(2,F)$ on $\mathbb{P}^1(F)$, for a field F, is:

$$\begin{pmatrix} a & b \\ c & d \end{pmatrix} \text{ sends } z \text{ to } \frac{az+b}{cz+d}\,;\ ad-bc \neq 0 \tag{6.10.5}$$

(*left action* is traditional here; the *right action* of a matrix in $GL(2,F)$ is gotten from the inverse matrix). Thus, L_1 is generated by the left actions of the matrices

$$x := \begin{pmatrix} 1 & 1 \\ 0 & 1 \end{pmatrix} \text{ and } t := \begin{pmatrix} 0 & 1 \\ -1 & 0 \end{pmatrix}. \tag{6.10.4}$$

(see (1.14)) which may be diagrammed as follows:

x : 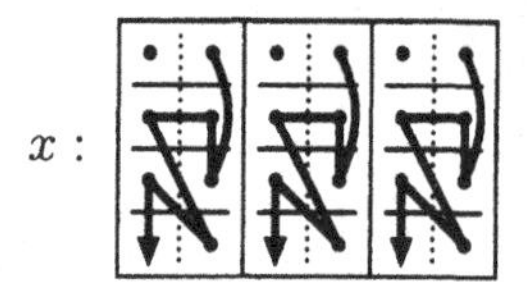 and t: 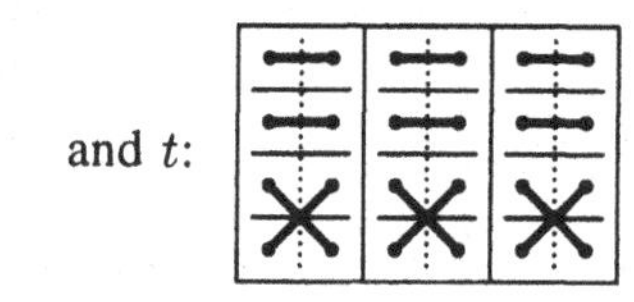

We argue that these permutations are in M_{24}. First, observe that our standard hexacode labeling $\mathcal{SL}_0$ (5.36) is carried by x^4 to the labeling $\mathcal{SL}_4$, whence $\langle x\rangle = \langle x^4\rangle \leq G$; with respect to $\mathcal{SL}_{4,1} t$ acts as the hexacode word $(1\omega|1\omega|1\omega)$, so is in M_{24}; see (5.38) where these are listed as UP1 and UP2; the fact that x is in M_{24} justifies the validity of the labelings $\mathcal{SL}_k$, $1 \leq k \leq 6$, given in (5.36). Since obviously x and t fix each octad $\mathcal{O}_i$, $\langle x,t\rangle \leq G(\mathcal{T})$.

The action of L_1 on R is faithful; for otherwise simplicity of L_1 implies that it would act trivially and so the action of R would leave invariant the fixed points of any element of L_1, which is certainly false; consider the action of x from (6.10.4), for instance.

We define L_2 as the group $\{r_g \mid g \in T\}$, where T is the natural Σ_3 in S; see (4.1), ff. Clearly, it commutes elementwise with L_1 and permutes the R_i as it does the $\mathcal{O}_i$ so acts faithfully on R (which equals $R_i \times R_j$, for any $i \neq j$). Trivially, we conclude that the action of $L_1 \times L_2$ is faithful since, by the Remak-Krull-Schmidt Theorem, a normal subgroup is generated by its intersections with L_1 and L_2. □

(6.11) Remark. Choose one of the $\mathcal{O}_i$ and consider the variation $L_{i,1}$ of L_1 generated by x and tu, where $u = c_i$ (so for $i = 3$, tu looks like

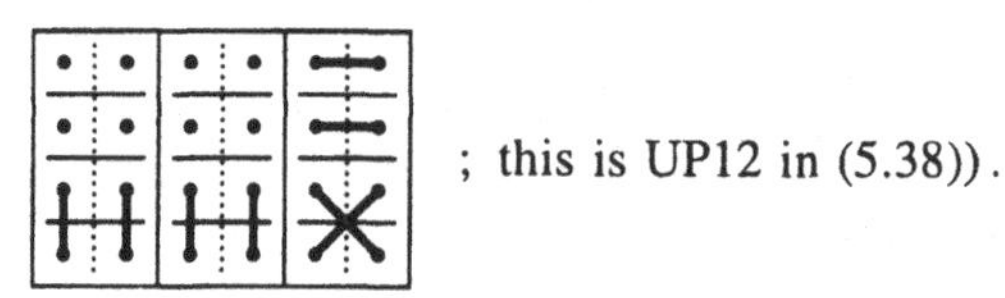

; this is UP12 in (5.38)).

Then, the actions of L_1 and $L_{i,1}$ on K_{12} are the same; let K be the kernel of this action. Then, $K \leq R(i)$, which is semiregular on $\Omega \backslash \mathcal{O}_i$. Since $L_{i,1}$ fixes a point in $\Omega \backslash \mathcal{O}_i$, so does K, whence $K = 1$. It follows that $L_{i,1} \cong PSL(2,7)$ and is a complement to $R(\mathcal{T})$ in $R(\mathcal{T})L_1$. We therefore have four conjugacy classes of complements to R in $R(\mathcal{T})L_1$. By (2.8), we see that there are at most four classes. Therefore, there are exactly four (and this also proves that $dim_{\mathbb{F}_2} H^1 \left(GL(3,2), \mathbb{F}_2^3\right) = 1$).

(6.12) Exercises

(6.12.1) Show that the group $G(\mathcal{O})$ is maximal. Hint: (6.9) indicates the possibilities for a block in a system of imprimitivity.

(6.12.2) Let $x \in G$ have order 7. Show that x has cycle shape $1^3 7^3$, centralizer of the form $7 \times \Sigma_3$ and normalizer of the form $7{:}3 \times \Sigma_3$.

(6.12.3) Show that the group $G(\mathcal{T})$ is maximal. Hint: be inspired by the strategy in (6.5.2), this time employing an element of order 7.

(6.12.4) Prove that the groups $R(\Xi), R(\mathcal{O})$ and $R(\mathcal{T})$ are weakly closed in a Sylow 2-group with respect to G. Hint: consider orbit structures. (Definition: if $A \leq B \leq C$ are groups, we say that A is *weakly closed in* B *with respect to* C if A is the only C-conjugate of A which lies in B.)

(6.13) Remark. There is another $PSL(2,7)$ whose existence follows from (6.27) by restricting from $\mathbb{F}_4$ to $\mathbb{F}_2$. Its orbit lengths are $1 + 1 + 1 + 7 + ?$, where $? = 7 + 7$ or 14; if $7 + 7$, the group stabilizes three disjoint octads, so is in a trio group. However, we know from (6.11) that this group would have to be in the conjugacy class of either L_1 or $L_{1,1}$, which is impossible from the orbit structure. Therefore, its orbit structure is $1 + 1 + 1 + 7 + 14$.

(6.14) Remark. If H is a group which acts transitively on sets X and Y, the actions are equivalent iff a stabilizer H_x, for $x \in X$, is conjugate to a point stabilizer H_y, for $y \in Y$. We use this fact several times to observe that actions of a group on two sets are inequivalent if there is a group element which has different orbit structures on the two sets.

(6.15) Lemma. (*i*) *M_{24} is transitive on dodecads.*

(ii) *If $\mathcal{D}$ is a dodecad, $Stab(\mathcal{D})$ is transitive on the ordered pairs of octads which sum to $\mathcal{D}$ and if $\mathcal{O}$ is an octad, $Stab(\mathcal{O})$ is transitive on the set of dodecads which meet $\mathcal{O}$ in a* 6*-set.*

(iii) *If $(\mathcal{D}, \mathcal{O})$ is such a dodecad-octad pair, its stabilizer in M_{24} is isomorphic to Σ_6, with orbits $6+6+10+2$; the orbits of length 6 afford inequivalent permutation representations and there is an element of M_{24} which interchanges them and induces an outer automorphism on this copy of Σ_6. For this Σ_6, a point stabilizer ($\cong \Sigma_5$) on one orbit of length 6 is transitive on the other orbit.*

Proof. (i) Let $\mathcal{D}$ and $\mathcal{D}'$ be dodecads. Write each as a sum of octads. Transitivity on octads allows us to write $\mathcal{D} = \mathcal{O} + \mathcal{O}'$ and $\mathcal{D}' = \mathcal{O} + \mathcal{O}''$. Using (6.9), we see that there is $g \in G(\mathcal{O})$ fixing $\mathcal{O}$ pointwise and carrying $\mathcal{O}'$ to $\mathcal{O}''$, and we are done.

(ii) Both statements follow from transitivity of G on the set of dodecad-octad pairs which meet in a 6-set, which in turn follows from (6.9) and transitivity on octads and on dodecads.

(iii) The number of such dodecad-octad pairs is $2576 \cdot \binom{12}{5} / \binom{6}{5}$ or $759 \cdot \binom{8}{2} \cdot 16$, and so the stabilizer H of a pair $(\mathcal{D}, \mathcal{O})$ has order 6! We show that H is isomorphic to Σ_6. Choose $\mathcal{O} = K_{12}$, $\mathcal{O}' = R_0 + K_6$ and $\mathcal{D} = \mathcal{O} + \mathcal{O}'$.

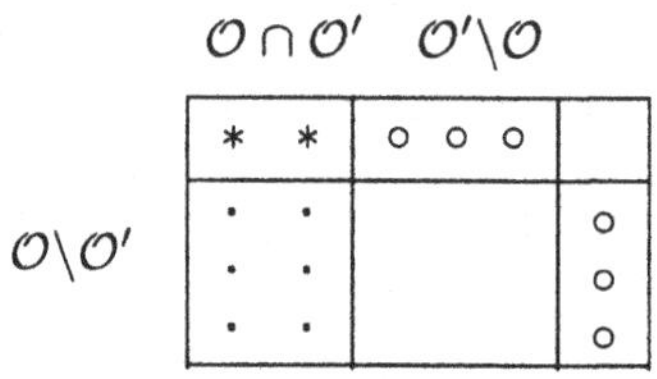

	$\mathcal{O} \cap \mathcal{O}'$	$\mathcal{O}' \backslash \mathcal{O}$	
	$*$ $*$	o o o	
$\mathcal{O} \backslash \mathcal{O}'$	· ·		o
	· ·		o
	· ·		o

We see that the permutations UP3 and UP4 stabilize all four subsets pictured. Thc permutation UP3 induces on 2-cycle on $\mathcal{O} \backslash \mathcal{O}'$ and a permutation of cycle shape 2^3 on $\mathcal{O}' \backslash \mathcal{O}$. The permutation UP4 induces a permutation of cycle shape $2^1 6^1$ on both. Also, an element of order 5 has cycle shape $5^4 1^4$ (see (6.12)). Hence an element of order 5 in H must induce a 5-cycle on each of the above 6-sets. The group Σ_6 is generated by any 2-, 6– and 5-cycle in it (an easy exercise since such a subgroup must be doubly transitive), so the natural map of H to $Sym\left(\mathcal{O} \backslash \mathcal{O}'\right) \cong \Sigma_6$ is an isomorphism onto. The first two of these elements induce on $\mathcal{O}' \backslash \mathcal{O}$ permutations of cycle shapes 2^3 and $3^1 2^1 1^1$, respectively, so the actions of H on these 6-sets are inequivalent. By (ii), there is an element $g \in M_{24}$ which stabilizes $\mathcal{D}$ and interchanges $\mathcal{O}$ and $\mathcal{O}'$, so normalizes H. Such an element does not induce an inner automorphism of H, or else we would have equivalent permutation representations of degree 6. Finally, we have to settle the orbit structure of H on the 2-set $X := \mathcal{O} \cap \mathcal{O}'$ and on $Y := \Omega \backslash \left(\mathcal{O} \cup \mathcal{O}'\right)$. An element $x \in H$ of order 5 has cycle structure $1^4 5^4$ on Ω hence 5^2 on Y. Since Σ_6 has no faithful permutation representation of degree 5 or less, H is transitive on Y. Since H acts faithfully on $\mathcal{O}$ and $G(\mathcal{O})$ maps onto $Alt_{\mathcal{O}}, H$ has orbits $6+2$ on $\mathcal{O}$, hence is transitive on X. Note also that $\langle UP3 \rangle$ acts transitively on X. $\square$

(6.16) Remarks. (i) If $(\mathcal{D}, \mathcal{O})$ satisfies (6.15.ii), so does $(\mathcal{D}, \mathcal{O} + \mathcal{D})$. (ii) The Σ_6 of (6.15.iii) lies in an octad stabilizer $2^4{:}GL(4,2)$, but is not conjugate to a natural Σ_6 in a natural $GL(4,2)$-subgroup, for otherwise it would fix a point. We remark that $dim\, H^1\left(Alt_6, 2^4\right) = 1$, $dim\, H^1\left(\Sigma_6, 2^4\right) = 1$ and $H^1\left(Alt_8, 2^4\right) = 0$; see (2.6.1), (2.16) and (2.18). (iii) We get a Steiner system with parameters (5,6,12) on a dodecad $\mathcal{D}$ by taking $\{\mathcal{D} \cap \mathcal{O} \mid \mathcal{D} \cap \mathcal{O}$ is a 6-set $\}$.

(6.17) Definition. Let M_{12} be the subgroup of M_{24} fixing a dodecad. It has order $2^6 3^3 5 \cdot 11$. We define M_{12-k} to be a point stablizer for the action of M_{12} on this dodecad, for $k \leq 5$.

(6.18) Theorem. (i) *M_{12} is sharply 5-transitive on the dodecad which defines it and on its complement (thus, the groups* (6.17) *are unique up to conjugacy);* (ii) *the actions of M_{12} on the stabilized dodecad and its complement are inequivalent. In particular, the group M_{11} defined as a point stabilizer in one dodecad is* 3*-transitive on the other dodecad, hence on Ω has orbit structure* $12 + 1 + 11$.

Proof. (i) It suffices to prove 5-transitivity since $|M_{12}| = 12 \cdot 11 \cdot 10 \cdot 9 \cdot 8$. Let F and F' be ordered 5-sets on the defining dodecad, $\mathcal{D}$. We get pairs $(\mathcal{D}, \mathcal{O})$, $(\mathcal{D}, \mathcal{O}')$ as in (6.15) and there is an element of G carrying one to the other. So, without loss, we may assume that $\mathcal{O} = \mathcal{O}'$ and so F and F' are ordered 5-sets within the 6-set $\mathcal{O} \cap \mathcal{D}$. We now use (6.15.iii). The last statement is immediate since we could have used the complement to define M_{12}.

(ii) Use (6.15.iii) for the first statement. For the 3-transitivity, note that on the second dodecad, our M_{11} does not fix a point and that an element of order 11 effects an 11-cycle on it (e.g., by (1.4); or look ahead to (6.21)). So, we have 2-transitivity. Let H be the stabilizer of two points; since it contains an element of order 5, the orbit decompostion on the other 10 points is $5+5$ or 10. We eliminate $5 + 5$, so assume that this is the case. Then, an element $x \in H$ of order 3 has at most one 3-cycle on each orbit so as an element of M_{12} has cycle structure $1^{12-3a}3^a$, for a = 1 or 2. Since x fixes 5 points, we contradict sharp 5-transitivity. □

(6.19) Exercises. (i) Suppose $h \in L(\Xi) \leq G(\Xi)$(6.3) has order 5. Prove that $C_{L(\Xi)}(h) \cong 3\times 5$, $C_{R(\Xi)}(h) \cong 2^2$ and that $C_{G(\Xi)}(h) \cong 5 \times Alt_4$ operates faithfully on the four orbits of length 5 of $\langle h \rangle$. Prove that $N_{G(\Xi)}(\langle h \rangle)/C_{G(\Xi)}(h) \cong 4$. Conclude that any two orbits of length 5 are contained in a unique dodecad and that if $\mathcal{D}$ is such a dodecad, $Stab(\mathcal{D}) \cap N_{G(\Xi)}(\langle h \rangle) \cong 5{:}4$, a Frobenius group of order 20.

(ii) Prove that Sylow 11-normalizers in M_{11}, M_{12} and Alt_{12} are of shape $11{:}5$. Conclude that M_{12} acts with even permutations only on the dodecad it stabilizes (we see in (6.24) that M_{12} is simple.)

(iii) Show that M_{11} has one conjugacy class of subgroups of index 12 (by (6.18), such a class exists). Hint: study a Sylow 11-normalizer (shape $11{:}5$), then a Sylow 5-normalizer (shape $5{:}4$; see part (i)). Notice that a group of shape $5{:}4$ has a unique subgroup isomorphic to $5{:}2$. A subgroup of index 12 is generated by subgroups of shape $11{:}5$ and $5{:}2$. (Such a subgroup is isomorphic to $PSL(2, 11)$, as we see in (6.21.i)).

(6.20) The PGL₂(11)-labeling

∞	∞	3	3	4	4
0	0	2	2	X	X
1	8	9	6	5	7
8	1	6	9	7	5

The dodecads L and R (left and right) pictured below are thus labeled with $\mathbb{P}^1(11)$, the projective line over $\mathbb{F}_{11}$:

R	L	R	L	R	L
L	R	L	R	L	R
L	R	L	R	L	R
L	R	L	R	L	R

The subgroup $PSL(2,11)$ of M_{24} operates by linear fractional transformations on both L and R. An outer automorphism of $PSL(2,11)$ is induced by the element τ of M_{24}, defined by $s_L \leftrightarrow s_R$, for $s \in \mathbb{P}^1(11)$; τ is UP2. The element of order $11\,(Y \mapsto Y+1\,mod\,11)$ is UP5; see (5.38).

(6.21) Notation. Consider the following permutations $\alpha, \beta, \ldots$ in M_{24} which stabilize the dodecads L and R, whose actions $\alpha_P, \beta_P, \ldots$ on $P \in \{L, R\}$ are given by:

$UP5$	$\alpha_L = \alpha_R^{-1} = (\infty)(0123456789X)$
$UP16$	$\beta_L = \beta_R = (\infty)(0)(13954)(267X8)$
$UP14$	$\gamma_L = \gamma_R = (\infty 0)(1X)(25)(37)(48)(69)$
$UP17$	$\delta_L = \delta_R = (\infty)(0)(1)(8)(34)(2X)(59)(67)$
$UP13$	$\tau'_L = \tau'_R = (\infty 0)(23)(4X)(18)(69)(57)$
$UP15$	$\gamma'_L = \gamma'_R = (\infty)(0)(14)(27)(35)(6)(8X)(9)$.

See (5.38); note that $\gamma = \gamma'\tau', \gamma'$ inverts β and τ' centralizes β.

(6.22) Proposition. (i) $\langle\alpha,\beta,\gamma\rangle = \langle\alpha,\gamma\rangle \cong PSL(2,11)$; (ii) $\langle\beta,\gamma\rangle \cong Dih_{10}$ *and has orbits* $2+10$ *on* $\mathcal{D}$ *and* $\Omega\backslash\mathcal{D}$; (iii) $\langle\beta,\gamma'\rangle \cong Dih_{10}$ *and has orbits* $1+1+5+5$ *on* $\mathcal{D}$ *and* $\Omega\backslash\mathcal{D}$; (iv) $\langle\gamma,\delta\rangle \cong 2^2$; (v) $\langle\beta,\gamma',\delta\rangle \cong Alt_5$; (vi) $\langle\beta,\gamma,\delta\rangle \cong 2 \times Alt_5$; (vii) $\langle\alpha,\beta,\gamma,\delta\rangle = M_{12}$; (viii) $\langle\alpha,\beta,\gamma'\rangle \cong PSL(2,11)$ *(this is the stabilizer in* M_{12} *of the points from* L *and* R *labeled with* ∞*).*

Proof. (i) Use the fact that the action of $PSL(2,p)$ on $\mathbb{P}^1(p)$ is generated by $z \mapsto z+1$ and $z \mapsto -1/z$. Next, (ii), (iii) and (iv) are straightforward. To get (v), define $T := \{\infty,0\}_L \cup \{\infty,0\}_R$ and let Y be the set of orbits of $\langle\tau,\tau'\rangle$ on $\Omega\backslash T$ and notice that $A := \langle\beta,\gamma',\delta\rangle$ preserves Y. Let $\varphi : A \to \Sigma_Y$. It is clear that $A^\varphi \le Alt_Y$ and since $\langle\beta\rangle^\varphi \ne \langle\beta^\delta\rangle^\varphi$ are distinct groups of order 5, $A^\varphi = Alt_Y$. We show that $Ker\,\varphi = 1$. Now, A leaves invariant the sextet Ξ' containing T (this is the sextet for $\mathcal{SL}_4$; see (5.38)) and $Ker\,\varphi$ leaves it partwise fixed. Suppose $1 \ne g \in Ker\,\varphi$; then g fixes T pointwise. On at least four parts of Ξ', g is fixed point free. Since g fixes L, it fixed all $L \cap T', T \in \Xi'$ and so we may assume that g has shape

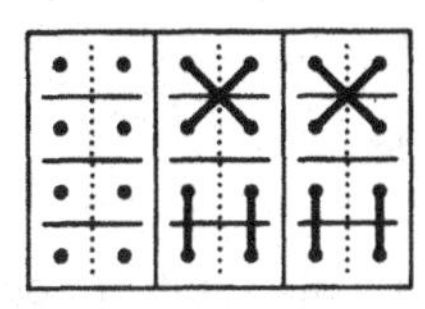

whence $g\tau$ has shape

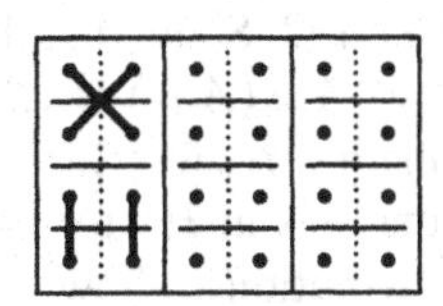

Since $g\tau$ fixes an octad pointwise and a point outside, it is the identity (6.8), a contradiction. So, (v) is proved.

We now derive (vi) from (v). Since γ and γ' induce the same permutation on the orbits of τ' and is nontrivial on $\{\infty 0\}_L$, we have an epimorphism from $\langle\beta, \gamma, \delta\rangle$ to Alt_5. Since $\gamma = \gamma'\tau'$, $\langle\beta, \gamma, \delta\rangle$ is a subgroup of $\langle\beta, \gamma', \delta\rangle \times \langle\tau'\rangle \cong Alt_5 \times 2$ and so $\langle\beta, \gamma, \delta\rangle = \langle\beta, \gamma', \delta\rangle \times \langle\tau'\rangle$, proving (vi).

Let $H := \langle\alpha, \beta, \gamma, \delta\rangle$. Then, H has index dividing 12^2 in M_{12} and contains $H_1 := \langle\alpha, \beta, \gamma\rangle \cong PSL(2, 11)$. Since H_1 has a single conjugacy class of involutions, represented by γ, and the involutions δ and γ have different cycle structures, we conclude that $H_1 < H$. From (6.19.ii), if $H < M_{12}$, its index is exactly 12. Notice that H has no subgroup of index 2; for, if K is such a subgroup, simplicity of H_1 implies that $H_1 \leq K$ and K has index not $1 (mod\, 11)$, a contradiction. Now, (1.13) implies that every involution of H is conjugate to an element of H_1, a contradiction.

To prove (viii), use (6.19.iii) and (6.18) to see that $\langle\alpha, \beta, \gamma'\rangle$ is a subgroup of index 12 in M_{11} (uniquely determined up to conjugacy). The isomorphism type may be verified by checking a presentation (1.27). Note that this group has orbits $1 + 11$ on each dodecad, so we cannot identify it by labeling with $\mathbb{P}^1(11)$. □

(6.23) Remarks. (i) Note that the involutions γ and δ are not conjugate in Σ_{24}. (ii) The element τ centralizes the subgroup $\langle\beta, \gamma, \delta\rangle \cong 2 \times Alt_5$ of M_{12} and satisfies $\alpha^\tau = \alpha^{-1}$. (iii) The groups (6.22.i) and (6.22.viii) are not conjugate in M_{24} (by (i) or the different orbit structures).

(6.24) Theorem. *M_{12} is simple.*

Proof. A proof along the lines of (5.33) can be given. Instead, we take a minimal normal subgroup, K. By primitivity, K is transitive; see (1.3.ii). So, $12 \mid |H|$. Let H be the above $PSL(2, 11)$-subgroup. If $H \cap K = 1$, KH is a subgroup of index dividing 12, so that K has order dividing 144, hence is solvable; but then a proper p-subgroup of K would be normal in M_{12}, a contradiction to K being minimal normal. So, $H \cap K \neq 1$. Since H is simple, $H \leq K$. Sylow theory for the prime 11 implies that K contains a Sylow 11-normalizer and so is self-normalizing, whence $K = M_{12}$. □

(6.25) Exercises

(6.25.1) Show that M_{11} is simple.

(6.25.2) Show that M_{10} is not simple (hint: it contains the group H' of (6.15)) and that it has orbits $12+1+1+10$ on Ω. Show that the stabilizer H^* of a 2-set in a dodecad stabilizer is isomorphic to $Aut\,(Alt_6) \cong Alt_6\,.2^2$ (use $Alt_6 \cong PSL\,(2,9)$ ([Hu], p. 193), then quote Steinberg's result [Car], Chapter 12). Show that M_{10} consists of the set of even permutations in H^* for its action on the length 10 orbit and that $M_{10}\backslash M_{10}'$ contains no involutions. Show that $M_9 \cong 3^2\!:\!Quat_8$. Hint: (1) In Alt_6, a Sylow 3-normalizer has shape $3^2\!:\!4$; now use the fact that $M_{10}\backslash M_{10}'$ contains no involutions. (2) Argue that M_9 is sharply doubly transitive on 9 points, is embedded in $AGL\,(2,3) \cong 3^2\!:\!GL\,(2,3)$, which has semidihedral of order 16 Sylow 2-subgroups, and has no element of order 8. (3) Look ahead to Section 7.

(6.25.3) This is more difficult. Show that M_{12} has precisely two conjugacy classes of subgroups of index 12 (both isomorphic to M_{11}) Hint: study the Sylow 11-normalizers and 5-normalizers in such a subgroup and compare with the situation for G (and compare with exercises (6.19) and (2.8)).

(6.25.4) Show that any $PSL\,(2,11)$ subgroup of G is conjugate to one given in (6.22.i) or (6.22.viii). (The hint for (6.25.3) is good here too).

(6.25.5) Show that, in the notation of (6.21), the centralizer of τ in M_{12} has shape $2 \times Alt_5$ and that the normalizer of this group in $M_{12}\langle\tau\rangle$ is of shape $\left[2^2 \times Alt_5\right] :>$ 2 (2.26). Show also that $C_{M_{12}}(\tau) \cong 2 \times \Sigma_3$.

(6.26) The PSL(2, 23) labeling

∞	0	22	1	11	2
3	15	12	21	13	7
6	5	18	20	4	10
9	19	8	14	16	17

We label Ω with $\mathbb{P}^1(23)$ as above. That this works is a straightforward exercise with (5.37). We comment that this group meets the $PSL\,(2,11)$ of (6.20) in the group $\langle\alpha\rangle$ of order 11 and it meets $PGL\,(2,11)$ in $\langle\alpha,\tau\rangle \cong Dih_{22}$. The Golay code is spanned by those dodecads which are the images under this $PSL\,(2,23)$ of $\{0,1,2,3,4,6,8,9,12,13,16,18\}$, the set of squares in $\mathbb{F}_{23}$. Thus, the Golay code is an example of a quadratic residue code [MacW-Sl], p. 481.

(6.27) The M_{21} labeling. The group M_{21} and its action on the 21 points not fixed by it are equivalent to the group $PSL\,(3,4)$ and its action on the projective plane $\mathbb{P}^2\,(4)$.

∞	0	00	10	$\omega 0$	$\bar\omega 0$
I	1	01	11	$\omega 1$	$\bar\omega 1$
II	ω	0ω	1ω	$\omega\omega$	$\bar\omega\omega$
III	$\bar\omega$	$0\bar\omega$	$1\bar\omega$	$\omega\bar\omega$	$\bar\omega\bar\omega$

mnemonic:

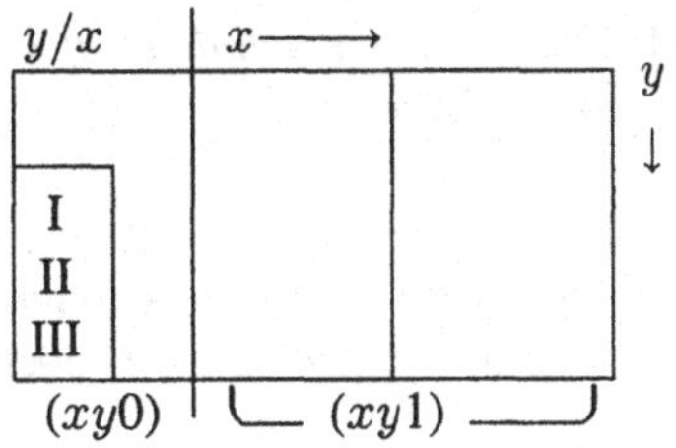

y/x	$x \longrightarrow$		$y \downarrow$
I II III			
$(xy0)$	$(xy1)$		

The notation is to be interpreted as follows. Three points are labeled with I, II and III (the *three Romans*) and the others are labeled with a point of $\mathcal{P}' := \mathbb{P}^2(4)$. Such a point is given by a representing nonzero vector (x, y, z) and such a point is denoted by the element α of the projective line $\mathbb{P}^1(4)$ when $z = 0$ and $\alpha = y/x$ by the pair (x, y) when the representing point satisfies $z = 1$.

(6.28) Notation. Let π be the set monomorphism $\mathcal{P}' \to \Omega$ defined as above. Let $\mathcal{R}$ (for Romans) denote the three exceptional points of Ω and let $\mathcal{P} := Im(\pi) = \Omega \backslash \mathcal{R}$.

(6.29) Remarks. (i) $Aut(\mathcal{P})$ is the group $P\Gamma L(3,4)$, the quotient of $\Gamma L(3,4)$, the group of invertible semilinear transformations, by the scalar transformations; see [Artin], (2.26).

(ii) Octads may be interpreted in the geometry of $\mathcal{P}$. The Leech triangle tells us that every subset of $\mathcal{R}$ (6.30) occurs as the intersection of $\mathcal{R}$ with an octad. Suppose $\mathcal{O}$ is an octad and that $\mathcal{O} \cap \mathcal{R}$ is a k-set.

$k = 0$: $\mathcal{O} = L + L'$, where L and L' are lines in $\mathcal{P}$;

$k = 1$: $\mathcal{O} \cap \mathcal{P}$ is a subplane over $\mathbb{F}_2$;

$k = 2$: $\mathcal{O} \cap \mathcal{P}$ is the image in $\mathcal{P}$ of a subset of $\mathbb{F}_4^3$ of the form $[Q^{-1}(0) \cup rad(V)] \setminus \{0\}$, where Q is a nondegenerate quadratic form on V; (3.2).

$k = 3$: $\mathcal{O} \cap \mathcal{P}$ is a line in $\mathcal{P}$.

We verify the statement for $k = 0$. It suffices to check the statement for a particular such octad (e.g. K_{56}), use (6.32) to get that M_{21} preserves the geometry of $\mathcal{P}$ and by (6.8) is transitive on the set of 210 octads which meet $\mathcal{R}$ in the empty set. To verify the case $k = 1$, study (6.8) and (6.10) and the subplane of points represented by $(x, y, z) \in \mathbb{F}_2^3$. For $k = 2$, use (6.15.iii), which implies that the stabilizer in M_{21} of such a 6-set is isomorphic to Alt_6 and the quadratic form $Q, Q(x, y, z) := x^2 + xy + y^2 + z^2$. The radical of the associated bilinear form is spanned by $(0, 0, 1)$ and the singular points are represented by $(a, b, 1)$, $a, b \in \mathbb{F}_2$, $(a, b) \neq (0, 0)$, $(1, \omega, 0)$ and $(1, \bar\omega, 0)$. Since $GL(3,4)$ is transitive on the set of nondegenerate quadratic forms, interpretation for $k = 2$ follows for all such intersections, once we prove that the stabilizer in M_{24} of $\{\mathrm{I}, \mathrm{II}, \mathrm{III}\}$ induces $PGL(3,4)$ on $\mathcal{P}$; see (6.30).

(6.30) Proposition. *For every element g of $P\Gamma L(3,4)$, there is a unique element $g' \in M_{24}$ such that, for each $p \in \mathcal{P}'$, $pg\pi = p\pi g'$. The assignment $g \mapsto g'$ is a group* *homomorphism which embeds* $Aut(\mathcal{P}')$ *as* H, *the subgroup of* M_{24} *stabilizing* $\{\mathrm{I}, \mathrm{II}, \mathrm{III}\}$. *The pointwise stabilizer of* $\{\mathrm{I}, \mathrm{II}, \mathrm{III}\}$ *is the group* M_{21} *and is the image of* $PSL(3,4)$ *under* π.

Proof. In the notation of Proposition (1.19), we show that (1) Σ_3 lifts to H; (2) diagonal matrices lift to H; (3) enough upper unipotent matrices lift to H; (3) a semilinear map associated to the Frobenius lifts to H.

Step (1). The generating transpositions (12) and (23) are represented by the respective permutations UP3 and UP4.

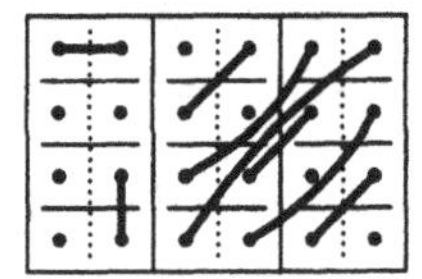
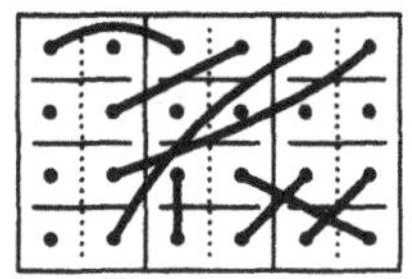

Step (2). The group of diagonal matrices in $GL(3,4)$ is generated modulo the center by the images of $diag(1, \omega, \bar{\omega})$ and $diag(\omega, 1, 1)$, which are represented by the respective permutations UP8 and UP9:

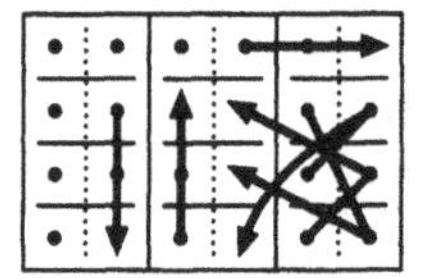
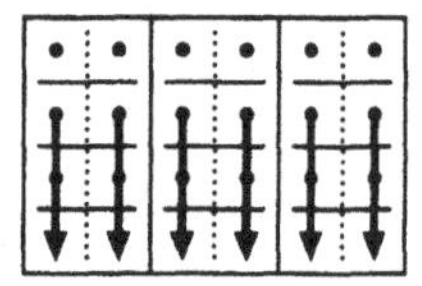

We use the "root group" (see (1.19)) generated by involutions $(x, y, z) \mapsto (x + cy, y, z)$, for $c = 1$ and ω, which are represented by permutations UP10 and UP11:

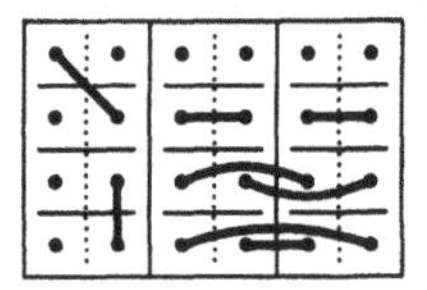
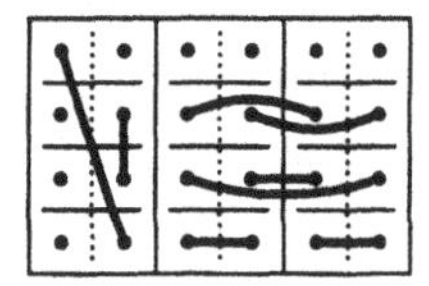

Step (3). A field automorphism: any of the following will do: (01****) acting via $\mathcal{SL}_3$, (01****) acting via $\mathcal{SL}_2$, (01****) acting via $\mathcal{SL}_4$; their partial cycle diagrams are below.

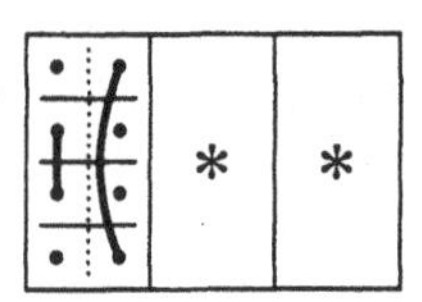
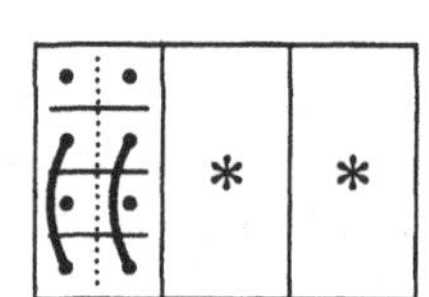
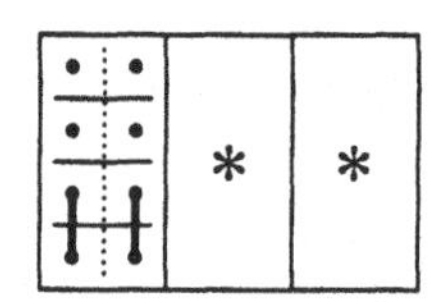

□

We now move on to study our last system of labeling; see (6.39).

(6.31) Definition. Let $\mathfrak{X}$ be a subset of $\{\Omega$, dodecad, octad, 16-sets$\}$. A linear subspace of $\mathcal{G}$ is called *$\mathfrak{X}$-pure* if any nonzero set in it has a type from $\mathfrak{X}$. (We may write "universe" for Ω.)

(6.32) Definition. A six-set in Ω is called *special* if it is contained in an octad; otherwise, it is called *transverse.* (From (6.8), we see that there is one M_{24}-orbit of each type).

(6.33) Lemma. Let D be a dodecad and X a 6-subset of D. Define $\mathcal{A} := \{A \in \mathcal{G} | A \cap D = X\}$. Then exactly one of the following is true (according to whether X is special or transverse):

(i) $\mathcal{A}$ consists of an octad and a 16-set;
(ii) $\mathcal{A}$ consists of two dodecads;

In these respective cases, the stabilizer of D and X is a Σ_6 subgroup with orbits $6+6+2+10$ and a Σ_5-subgroup with orbits $6+6$ on D and 12 on $D+\Omega$.

The stabilizer of $\{X, X+D\}$ in these respective cases is $Aut(\Sigma_6)$ (orbits $12+2+10$) and $\mathbb{Z}_2 \times \Sigma_5$ (orbits $12+12$). In the latter case, the stabilizer H of the partition $\Pi := \{X, X+D, X+A, \Omega+X+A\}$ has the form $H = (Alt_4 \times Alt_5) :> 2$ (see (2.26) for this notation), and induces full symmetric group on Π.

Proof. Let $p: \mathcal{G} \to P(D)$, $A \mapsto A \cap D$. Then $Ker\, p = \{0, D\}$, and $Im\, p = PE(D)$. It follows that $\mathcal{A}$ is a coset of $Ker\, p$ (and so is nonempty). If X is contained in an octad, we clearly are led to (i) and in the other case, if $A \in \mathcal{A}$, A is not an octad nor a 16-set (or else $A+D+\Omega$ would be an octad containing X), so we are led to (ii).

See (6.14.iii) for the stabilizing subgroup in case (i). Suppose case (ii) applies. Let $L = Stab(D) \cap Stab(X)$. Notice that, if Y is any 5-set in X and $\mathcal{O}(Y)$ is the octad containing it, $Y^* := \mathcal{O}(Y) \cap D$ is a 6-set meeting X in Y and that, as Y varies, the sixth point of $\mathcal{O}(Y) \cap D$ ranges over a 5-set Y' of $D \backslash X$. Thus, the stabilizing group J of Y is a Σ_5-subgroup of $Stab(D) \cap Stab(Y^*)$. This group J is transitive on $D+Y^*$ (see (6.9.iv)), so the stabilizer of X in J is a group $K \cong Frob(20)$. The orbits of K are $1+5+1+5$ on D.

Since the above applies to any choice of Y in X, we conclude that L induces a transitive group on X. The stabilizer in L of a point of X is the group $K \cong Frob(20)$, which does act faithfully, so the action of L on X is faithful and $L \cong \Sigma_5$ (6.25.5).

We now analyze the orbits of L on Ω. Within D, X and $X+D$ are the orbits. Within $D+\Omega$, we have orbits $2+10$ for K. As Y varies within X, so does the 2-set $\mathcal{O}(Y) \cap (D+\Omega)$; furthermore, these 2-sets partition $D+\Omega$, else we have distict octads meeting in 5 points. Therefore, L is transitive on $D+\Omega$.

Now to prove the final statements about the stabilizers of $\{X, X+D\}$. These groups are twice as large since transitivity on pairs (X, D) gives an element of G which carries (X, D) to $(X+D, D)$. For the structure of these groups, see (6.14) for case (i) and for case (ii), we refer to (6.20); we note that the $\mathbb{Z}_2$-direct factor in case (ii) is generated by an involution with cycle structure (2^{12}) and that its orbits on $D + \Omega$ form a system of imprimitivity for the action of the group L.

Finally, take the partition Π as in the statement of (ii). The sum of any two members of Π is a dodecad and we may then construct an element of H inducing a transposition on Π fixing those two members. The subgroup of L fixing each member of Π is $L' \cong Alt_5$. Since H contains $L \cong \Sigma_5$, we get $H = (Alt_4 \times Alt_5) :> 2$ (2.26). □

(6.34) Lemma. *Let S be a dodecad-pure subspace of $\mathcal{G}$ which has dimension* 2.

(i) *There is an element $g \in G(\Xi)$ such that S^g is generated by*

$$(6.34.1) \qquad D_0 := \left|\begin{array}{cc|cc|cc} & & & & & \\ \circ & \circ & \circ & \circ & \circ & \circ \\ & \circ & & \circ & & \circ \\ \circ & & \circ & & \circ & \end{array}\right|, \qquad E_0 := \left|\begin{array}{cc|cc|cc} & & & & & \\ \circ & & \circ & & \circ & \\ \circ & \circ & \circ & \circ & \circ & \circ \\ & \circ & & \circ & & \circ \end{array}\right|.$$

(ii) *If T is another such space whose every member avoids R_0 (the top row) and if $S \cap T \neq 0$, then $S = T$.*

Proof. Let (D, E) be a basis of S and set $X = \Omega \setminus [D \cup E]$. Since X is a transverse 6-set, there is $g \in G$ such that $X^g = R_0$. So, we assume $X = R_0$ and observe that every set in S has even parity. Let $A \in S$ and consider $\mathcal{L}(A)$. Since A is a 12-set and each $A \cap K_i$ is even, each $A \cap K_i$ is a 2-set, whence $\mathcal{L}(A)$ is a weight 6 hexacode word. From (4.7), we know that there is a single orbit of 2-dimensional pure weight 6 subspaces of $\mathcal{H}$ under $Aut^*(\mathcal{H})$; such 2-spaces are actually $\mathbb{F}_4$-subspaces. Consequently, (i) and (ii) follow at once. □

(6.35) Corollary. *M_{24} operates transitively on the set of 3-dimensional dodecad-universe pure subspaces which contain Ω. The stabilizer of such a subspace is a group of the form $(Alt_4 \times Alt_5) :> 2$ (see (2.26) for this notation); the subspace is an idecomposable module with ascending factors of dimensions* 1 *and* 2.

Proof. Let S be such a subspace and let $D, E \in S$ be dodecads which are independent modulo $\langle\Omega\rangle$. Then $S = \langle D, E\rangle \oplus \langle\Omega\rangle$. Now use (6.33) and (6.34.i). □

(6.36) Lemma. *Let D, E and F be linearly independent dodecads in a dodecad-pure space S in $\mathcal{G}$. Then, $|D \cap E| = 6$ and $|D \cap E \cap F| = 3$. Also, $\dim S = 3$.*

Proof. Obviously, $|D \cap E| = 6$. Let a, b, c, d be the cardinalities of the four intersections $F \cap X$, where $X = D' \cap E'$, and where D' denotes D or $D + \Omega$ and E' denotes E or $E + \Omega$. We have

$$a + b = 6$$
$$a + c = 6$$
$$b + c = 6,$$

which implies that $a = b = c = 3$, whence also $d = 3$. The easiest way to see that $dim\, S = 3$ is to transform the transverse 6-set $D \cap E$ to R_0 and notice that what we have proven is that, in the standard picture, $span\,\{D, E\}$ consists of even parity sets and that any $F \in S \setminus span\,\{D, E\}$ has odd parity. □

(6.37) Lemma. *Let U be a subspace of $\mathcal{G}$ which contains Ω.*

(i) *Suppose that $W = \{A \in U | p(A) = 0\}$ (p = parity) has codimension 1, $W^{\#}$ consists of dodecads and $W \cap \mathcal{L}^{-1}(0) = 0$. Let $d + 2 = dim\, U$. Then the stabilizer of U in G contains an elementary abelian 2-group $R \cong 2^d$ such that R is trivial on $W/\langle\Omega\rangle$ (and of course on $\langle\Omega\rangle$ and U/W), acts faithfully on $U/\langle\Omega\rangle$ and is faithful on W iff $d \geq 2$.*

(ii) *Suppose that $dim\, U \geq 4$, $U \setminus \langle\Omega\rangle$ consists of dodecads and that, for all subspaces P of codimension 1 in U and containing Ω, that there is $g \in G$ with $P^g = S^g \cap p^{-1}(0)$ and $P^g \cap \mathcal{L}^{-1}(0) = 0$. Then, $Stab\,(S)$ induces on $S/\langle\Omega\rangle$ the group $SL\left(S/\langle\Omega\rangle\right)$. Also, the module extension $0 \to \langle\Omega\rangle \to S \to S/\langle\Omega\rangle \to 0$ is nonsplit.*

Proof. (i) The restriction of $\mathcal{L}$ to W is a monomorphism and so we define $R := [\mathcal{L}(W)]^{\pi_1}$ (see (5.22)). This group has order 2^d and is in $G(\Xi)$. It is faithful on $U/\langle\Omega\rangle$ by (5.23.i) and is faithful on W if $d \geq 2$ since for any two elements u and v of W, which are independent modulo $\langle\Omega\rangle$, $\mathcal{L}(u)$, $\mathcal{L}(v)$ and $\mathcal{L}(u+v)$ all have weight 6; this means that $u \cap v$ is a 6-set transverse to the K_i and that the action of v^{π_1} carries u to $u + \Omega$. If $d = 1$, the image of $u \in W$ under u^{π_1} is just u, so the action on W is not faithful.

(ii) The group $SL\,(m, q)$ is generated by its root groups; see (1.19). Our hypotheses imply that, for any pair (u, H), where u is a nonzero vector of S and H is a hyperplane containing u, $Stab\,(S)$ contains the linear map $v \mapsto v + \varepsilon u$, where $\varepsilon = 0$ or 1 as $v \in H$ or $v \notin H$, so we get the action of $SL\left(U/\langle\Omega\rangle\right)$. The nonsplitting follows from the fact that $\Omega = A + A^g$, for some $g \in Stab\,(S)$ and some $A \in S$, for instance $g = v^{\pi_1}$ and $A = u$ as in the proof of (i). □

(6.38) Theorem. *Let S be a maximal dodecad-universe pure subspace. Then $\Omega \in S$ and $dim\, S = 4$. On the set of maximal dodecad pure subspaces, $G = M_{24}$ has two orbits, represented by, say, S^+ and S^-, where $S^\varepsilon = span\,\{D_0, E_0, F_\varepsilon, \Omega\}$ and where*

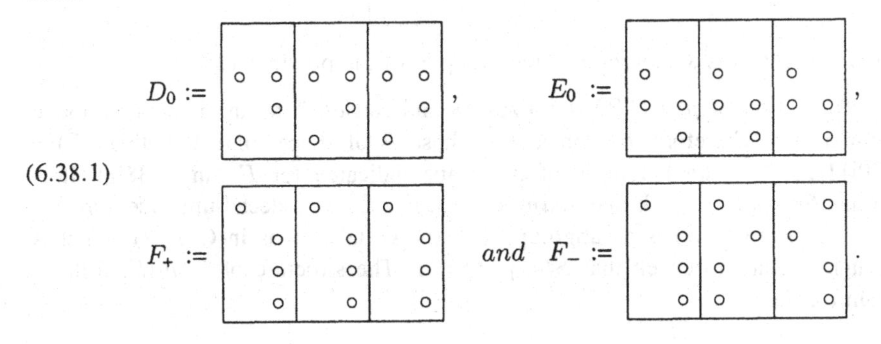

Both codes are thickened, extended Hamming codes (3.19), (3.14.b). *The stabilizers are $G(S^+)$, which is conjugate to the natural $\Sigma_3 \times L_2(7)$ subgroup of the trio group described in* (6.8), *and a transitive subgroup $G(S^-) \cong L_2(7)$ (which is not contained any proper subgroup of G described so far).*

Proof. Let S be a maximal dodecad-universe pure subspace of $\mathcal{G}$. By (6.34), we may assume that S contains D_0 and E_0. By (6.36), we know that $S_0 := S \cap p^{-1}(0) = span\{D_0, E_0, \Omega\}$ has codimension at most 1. Let $F \in S \backslash S_0$ (assuming that there is such an F). Let $H := Stab_{G(\Xi)}(S_0)$; by (4.6.2), $H \cong 3:\Sigma_5$ and modulo $O_3(H)$, is sharply 3-transitive on the set of six coordinate spaces. We use the notation $C(S)$ for $\{g \in Stab(S) \mid g \text{ is trivial on } S\}$.

Case 1. There exists such an F with $\mathcal{L}(F) = 0$ (i.e. $dim\,\mathcal{L}(S) = 2$). Using the action of H on R_0, we may assume that $R_0 \cap F = R_0 \cap F_+$, whence $F = F_+$. One verifies that every element in $span\{D_0, E_0, F_+, \Omega\}$ is a dodecad or Ω by simply checking the property in $F_+ + \langle E_0 \rangle$, then using the action of $\mu = UP9$ and taking complements. The centralizer of S_0 is contained in H, and the subgroup of H fixing the 3-set $R_0 \cap F$ is isomorphic to $3:\Sigma_3$ and the subgroup which furthermore fixes D_0 or E_0 is a Σ_3-subgroup, which is nothing other than the group T described in (4.1). Since this T plainly centralizes S, we have $C(S) = T$. By (6.37), we see that $Stab(S)$ induces $GL(3,2)$ on $S/\langle \Omega \rangle$. Let Q be the kernel of the action of $Stab(S)$ on $S/\langle \Omega \rangle$; then T is normal in Q. On S, $Stab(S)$ induces a subgroup of $2^3:GL(3,2)$ isomorphic to $GL(3,2)$ or $2^3:GL(3,2)$ (since the only possibilities for Q are 1 or 2^3). On the other hand, since Q fixes R_0 and S_0, it embeds into $3:\Sigma_5$ (4.6.3), which has Sylow 2-groups isomorphic to Dih_8. Therefore, $Q/T = 1$. We conclude that $Stab(S) \cong GL(3,2) \times \Sigma_3$ since T is normal in $Stab(S)$, $Stab(S)/T \cong GL(3,2)$ and $Aut(T) \cong Inn(T)$.

Case 2. There does not exist an F with $\mathcal{L}(F) = 0$ (i.e. $dim\,\mathcal{L}(S) = 3$). By (4.6.3.2), we may assume that $\mathcal{L}(S)$ has weight 4. Using the action of H, we may assume by (4.6.2) that $\mathcal{L}(F) = (00|11|11)$ and furthermore that F has the form

(6.38.2)

and that $C(S)$ is a subgroup of the group T of the previous case.

The action of $\mu = UP9$ stabilizes S_0 and carries F to an element as above with $\alpha = 1$. Therefore, by considering those F of shape (6.38.2) with $\mathcal{L}(F) = (00|11|11)$, we are led to F of the shape indicated for F_- in (6.38.1). It is straightforward to check that $span\{D_0, E_0, F_-, \Omega\}$ is dodecad-universe pure.

The group $C(S)$ is a subgroup of T (if we take F as in Case 2) and it is straightforward to check that $Stab_T(S) = 1$. The structure of $Stab(S)$ follows from (6.37).

We prove that $Stab(S)$ is transitive. First, notice that it contains a fours group whose nonidentity elements have cycle shape 2^{12}. If $Stab(S)$ were not transitive, the cycle structure $1^3 7^3$ for an element of order 7 implies that its orbit structure is $8+8+8$.

We finish by producing a space of this type and an $L_2(7)$-subgroup of G which stabilizes it and is transitive: see (6.40). □

(6.39) The $L_2(7)$-labeling

10 51	30 62	20 43
11 01	52 32	63 33
31 61	22 12	53 03
41 21	42 02	13 23

$= (\mathbb{F}_7^2 \backslash \{0\}) / \{\pm 1\}$.

The permutation x of (6.10) acts here as $\begin{pmatrix} 1 & 1 \\ 0 & 1 \end{pmatrix}$ and it is straightforward to check that UP18 acts as $\begin{pmatrix} 0 & 1 \\ -1 & 0 \end{pmatrix}$. Notice that $\begin{pmatrix} 3 & 0 \\ 0 & 2 \end{pmatrix}$ acts as the element $\mu^2 r_{(135)(246)}$. They are diagrammed below:

(6.39.1) 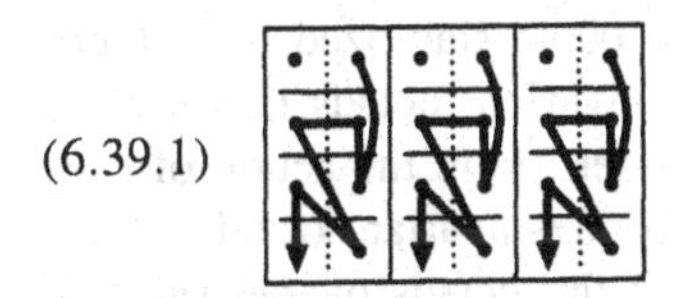 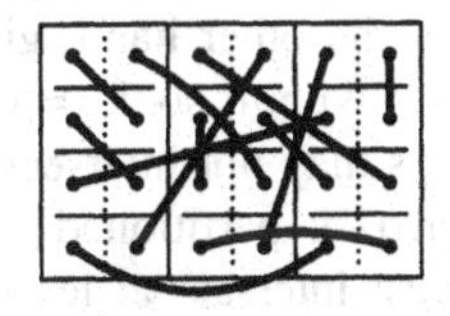 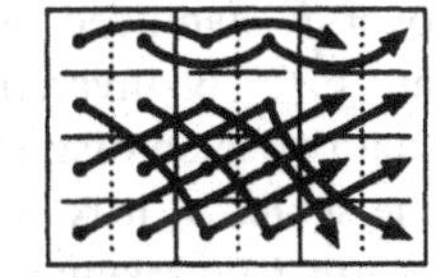

In (5.39), these are called UP1, UP18 and UP19.

(6.40) Remarks. (i) For $G(S^-)$, $P\Omega$ is the sum of two copies of the 8-dimensional Steinberg module and the projective indecomposable with the trivial module on top; also, $\mathcal{G}$ is the module direct sum of S^- and a Steinberg module. For $G(S^+)'' \cong GL(3,2)$, $P\Omega$ is the direct sum of three 8-dimensional projective indecomposables which cover the trivial module and $\mathcal{G}$ decomposes as $S^+ \oplus [\mathcal{G}, O_3(T)]$.

(ii) There is an essentially unique degree 24 transitive permutation representation of $PSL(2,7)$, hence the action of the transitive group $G(S^-)$ on Ω is equivalent to the action of $PSL(2,7)$ on $[\mathbb{F}_7^2 \backslash \{0\}] / \{\pm 1\}$.

We now prove a structure result about M_{24}.

(6.41) Theorem. *G has two conjugacy classes of involutions, of cycle shapes $1^8 2^8$ and 2^{12}. They are represented by elements in $R(\Xi)$ and correspond to, respectively, hexacode words of weight 4 and 6. Their centralizers have respective shapes $2^4{:}[2^3{:}GL(3,2)] = 2^{1+6}{:}GL(3,2)$ and $2^6{:}\Sigma_5$ and lie in, respectively, the group $G(\mathcal{O})$, where $\mathcal{O}$ is the octad of fixed points, and $G(\Xi)$.*

Proof. Let t be an involution with nonempty fixed point set, F. Since G is simple, every permutation is even and so $|F|$ is divisible by 4. Suppose $|F| > 4$. If E

is a 5-set in F and $\mathcal{O}$ the octad containing E, t acts on $\mathcal{O}$ with at least 5 fixed points, and so acts trivially, by (6.9). Therefore, $t \in R(\mathcal{O})$ and so is semiregular on $\mathcal{O} + \Omega$. All statements about t and $C(t)$ are now clear since t is trivial on $\mathcal{O}$ and acts as a translation in the action of $G(\mathcal{O})$ as $AGL(4,2)$.

If F is a 4-set, we argue as follows to get a contradiction.

Let A and B be any two distinct orbits of t of length 2. The octad containing A, B and a point of F is t-invariant. On it, t effects an even permutation of order 2 with fixed points, which therefore must have form $1^4 2^2$. It follows that such an octad must contain F. However, there are $\binom{10}{2} = 45$ choices of 4-set $A + B$, whereas F lies in only 5 octads, contradiction.

Now we assume that $F = \emptyset$, i.e., that t has cycle shape 2^{12}. We claim that there exists a sextet Ξ such that $t \in R(\Xi)$. Notice that $\mathcal{V}$, the set of 4-sets stable under t, has $\binom{12}{2} = 66$ members. Supposing the claim is false, we notice that, for each $F \in \mathcal{V}, t$ fixes the sextet $S = S(F)$ containing F, so must induce a nontrivial permutation of order 2 on the parts of S. The permutation must be even, or else t would effect an odd permutation on those parts it fixes (5.25); but t has no fixed points on Ω. Evenness and $Ft = F$ imply that t acts with cycle shape $1^2 2^2$ on the parts of S. Thus, there is a unique part $E \neq F$ of S stabilized by t, and of course $E + F$ is an octad. This proves that, given $F \in \mathcal{V}$, there is a unique octad over F stabilized by t. On the other hand, given an octad stabilized by t, there are $\binom{4}{2} = 6$ t-invariant 4-sets within, so $11 = 66/6$ t-invariant octads in all. Note that two such distinct octads may not intersect in a 4-set. From the action of t on the sextet S, it is clear that Ω is partitioned by a trio of t-invariant octads. But a fourth t-invariant octad must intersect at least one of the octads of the trio in a four-set, contradiction.

From the claim just proven, we have that $t \in R(\Xi)$, for some sextet Ξ. Thus, $t = r_h$, for some hexacode word of weight 6. We have $C := C_{G(\Xi)}(t) \cong 2^6 : \Sigma_5$. It remains to show that $C(t) \leq G(\Xi)$. This will follow from the statement that t leaves partwise invariant a unique sextet (i.e., $t \in R(\Xi) \cap R(\Xi')$ implies $\Xi = \Xi'$), which we now prove. Let S be another such sextet. Then, for every $K \in \Xi$ and $T \in S, K \cap T$ is a 2-set or is empty. We may use the action of $G(\Xi)$ to get a part of S to be

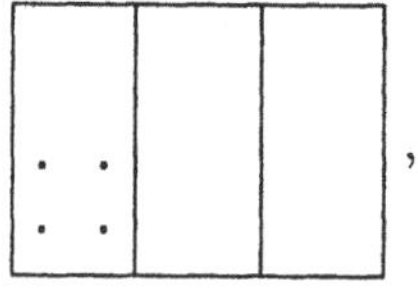

which implies at once that S looks like

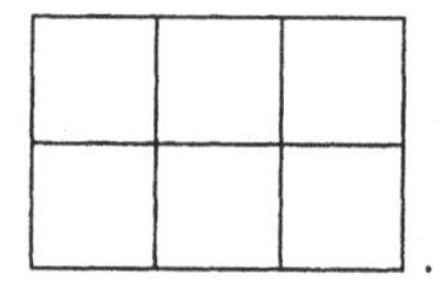

For the action of t to preserve this partwise, h must be a hexacode word with coordinates only 0 and 1, but since h has weight 6, this is a contradiction. □

(6.42) Exercises (6.42.1). Prove that M_{12} has two conjugacy classes of involutions.

(6.42.2) Show that M_{11} is not contained in M_{22} (hint: the Sylow 3-normalizers are incompatible). It is interesting to note that M_{22} has a character which looks as though it ought to be a permutation character for a subgroup of order $|M_{11}|$, but there is no such subgroup (this is a little harder to prove).

(6.42.3) Find Sylow 3-groups for each of the Mathieu groups by exhibiting permutations. Prove that M_{24} has two classes of elements of order 3 (see also (7.35)). Show that their normalizers have shape $3 \cdot \Sigma_6$ and $\Sigma_3 \times PSL(2,7)$.

(6.42.4) The groups $PSL(m,q)$ are simple if $(m,q) \neq (2,2)$ or (2,3). Use simplicity of $M_{21} \cong PSL(3,4)$ to prove simplicity of M_{22} and then M_{23}. Give a simplicity proof of M_{23} in the style of (5.33) (for M_{22}, this would be harder, due to the lack of a big prime).

(6.42.5) Prove that $PSL(3,2^n)$ has one conjugacy class of involutions. Hint: Jordan canonical form.

(6.42.6) Prove that $M_{21} \cong PSL(3,4)$, M_{22} and M_{23} have one conjugacy class of involutions. Hint: (6.42.5) and (1.13).

(6.43) Exercise. Show that linear subspace S of $\mathcal{G}$ which is maximal with respect to containing no dodecad is one of the following subspaces.

(6.43.1) For a fixed octad, $\mathcal{O}, S = \{A \in \mathcal{G} \mid A \cap \mathcal{O} = \mathcal{O} \text{ or } \emptyset\}$, a 6-dimensional space.

(6.43.2) For a sextet Ξ, $S = \{A \in \mathcal{G} \mid A \text{ is the sum of evenly many parts of } \Xi\}$, a 5-dimensional space.

(6.44) Exercise. Prove that the 2-rank of $GL(n,2)$ is $\left[\frac{n}{2}\right]\left(\left[\frac{n}{2}\right]-1\right)$.

(6.45) Exercise. Prove that the 2-rank of M_{24} is 6.

(6.46) Exercise. Prove that the 2-rank of M_{12} is 3.

(6.47) Exercise. Prove that if $t \in M_{24}$ has cycle shape $1^8 2^8$, then $dim\, C_{\mathcal{G}}(t) = 8$. Hint: t inverts an element of order 5.

Chapter 7. The Ternary Golay Code and $2{\cdot}M_{12}$

We give a brief treatment of the ternary Golay code, which leads to the group $2{\cdot}M_{12}$, the unique nonsplit perfect central extension of M_{12} (2.14). In particular, we prove existence and uniqueness results for the ternary Golay in the same sprit as that of Section 5.

The group $2{\cdot}M_{12}$ is not in M_{24}, though is easy to construct abstractly from M_{12} [Gr72a, 87a], p. 198, (I). Most information we need about M_{12} can be deduced from earlier sections, but the treatment in this section is useful for later work with the Leech lattice.

See Section 3 for code theoretic terminology.

(7.1) Definition. The *standard tetracode* is $\mathcal{T} := \{(s, a, a+s, a+2s)\,|\,a, s \in \mathbb{F}_3\}$. It is a ternary code with parameters $[4,2,3]$ and is self-orthogonal. We call s the *slope*.

A *tetracode* is a $[4,2,3]$ ternary code.

(7.2) Proposition. *There is a unique (up to equivalence) ternary code with parameters* $[4,2,3]$. *It is self-orthogonal, whence every weight is divisible by* 3. *The number of such codes is* $\binom{4}{3}2^3{\cdot}\binom{3}{2}{\cdot}2{\cdot}1{\cdot}2/8{\cdot}6 = 2^7 3/2^4 3 = 8$ *and so the automorphism group has order* $2^4 4!/8 = 2^4 3 = 48$.

Proof. Let C be such a code in $V := \mathbb{F}_3^4$; all nonzero weights are 3 or 4. If x and y are linearly independent vectors in C, they form a basis and some linear combination has weight 3. Let $x = (abc0)$ be such an element (reindexing if necessary). Then $y = (pqrs)$ may be arranged to satisfy $p = 0$ by subtracting a multiple of x; then $qrs \neq 0$. If $x.y \neq 0$, (bc) and (qr) are dependent vectors in $\mathbb{F}_3^2$ and $x - y$ or $x + y$ has weight 2, a contradiction. Therefore, $x.y = 0$ (whence C is self-orthogonal) and $(bc) = \pm(qr)$. A diagonal transformation now takes C to the standard tetracode. The last statment follows from analysis of the predeeding arguments. $\square$

(7.3) Lemma. *$Aut(\mathcal{T}) \cong GL(2,3)$ and it induces Σ_4 on the set of four columns.*

Proof. Direct calculation shows that the monomial transformations

$$\sigma = \begin{pmatrix} -1&0&0&0\\ 0&0&1&0\\ 0&1&0&0\\ 0&0&0&1 \end{pmatrix}, \quad \tau = \begin{pmatrix} -1&0&0&0\\ 0&0&0&1\\ 0&0&1&0\\ 0&1&0&0 \end{pmatrix}, \quad \rho = \begin{pmatrix} 0&0&-1&0\\ 0&0&0&1\\ 1&0&0&0\\ 0&-1&0&0 \end{pmatrix} \tag{7.3.1}$$

preserve $\mathcal{T}$. The subgroup of $GL(\mathcal{T})$ which they induce contains a Σ_3-subgroup and an element whose square is -1, whence is all of $GL(\mathcal{T}) \cong GL(2,3) \cong Quat_8{:}\Sigma_3$. From (7.2), we deduce that $Aut(\mathcal{T}) \cong GL(2,3)$, from order considerations. The last statement follows from (7.3.1) and the fact that Σ_4 is generated by permutations (23), (24) and (13)(24). □

(7.4) Exercise *(error correcting properties of the tetracode).* These are easy to prove, given the slope concept (7.1).

(7.4.1) Show that two given coordinate values extend to a unique tetracode word.

(7.4.2) Show that, given four coordinates, there is a unique tetracode word which agrees in at least two of these coordinates (uniqueness is clear since the minimum weight is three). In fact, such a tetracode word agrees in at least three coordinates. *Hint:* Suppose that $(c_\infty, c_0, c_1, c_2)$ is given. Let $d_i := c_{i+1} - c_i$, for $i = 0,1,2$ (indices read modulo 3). Then $d_1 + d_2 + d_3 = 0$. If the d_i are equal, correct c_∞ to d_i if necessary. If they are not all equal, $\{d_1, d_2, d_3\} = \{0,+,-\}$ and c_∞ equals one of them, say d_k; then correct c_{k-1}. Note that we have corrected at most one coordinate in either case.

(7.5) Notation. We let Δ be a 12-set and arrange it in a 3×4 array. We label the points of Δ by $l{:}\,\Delta \to \mathbb{F}_3 = \{0,+,-\}$, with the rows being the level sets. Let $K_1, \ldots, K_4$ be the columns. For indices $i, j, \ldots$, let $K_{ij\ldots}$ denote $K_i \cup K_j \cup \ldots$. A *labeled set* is a function from Δ to $\mathbb{F}_3$; it is essentially a 12-tuple, indexed by a pair c, i with $c \in \mathbb{F}_3$ and $i \in \{1,2,3,4\}$. When notating a labeled set with a 3×4 array, we typically write in only the labels + and −. If $x = (x_{ci})$ is a labeled set, we define the 4-*tuple labeling*, a homomorphism $\mathcal{L}$ from $\mathbb{F}_3^{\Delta}$ to $\mathbb{F}_3^4$ defined by

$$\mathcal{L}(x)_i = \sum_c c x_{ci}\,. \tag{7.5.1}$$

For example,

$$\mathcal{L}\left(\begin{array}{|c|ccc|} - & & + & + \\ & & + & - \\ - & & - & - \end{array}\right) = (+0-0)\,. \tag{7.5.2}$$

We will tolerate some confusion between the support of such a labeled set (i.e., the subset getting labels + and −) with the labeled set itself. We define a *signed point* to be a 1-set in Δ labeled with +; the set of signed points is denoted $\widetilde{\Delta}$.

(7.6) Notation. For each $i = 1,2,3,4$, define the *labled column* C_i as the labeled set which gets + on K_i and zeroes elsewhere. For $t \in \mathbb{F}_3^6$, define the *associated labeled set* (t) by $(t)_{c,i} = +$ iff $t_i = c$ and $= 0$ otherwise. For instance,

(7.6.1)

$$((00+-)) = \begin{array}{c|c|ccc|} 0 & + & + & & \\ + & & & + & \\ - & & & & + \end{array} \, .$$

The map $t \mapsto (t)$ is therefore just a section for the map $\mathcal{L}$, For the 9 tetracode words, the associated labeled sets are:

(7.6.2)

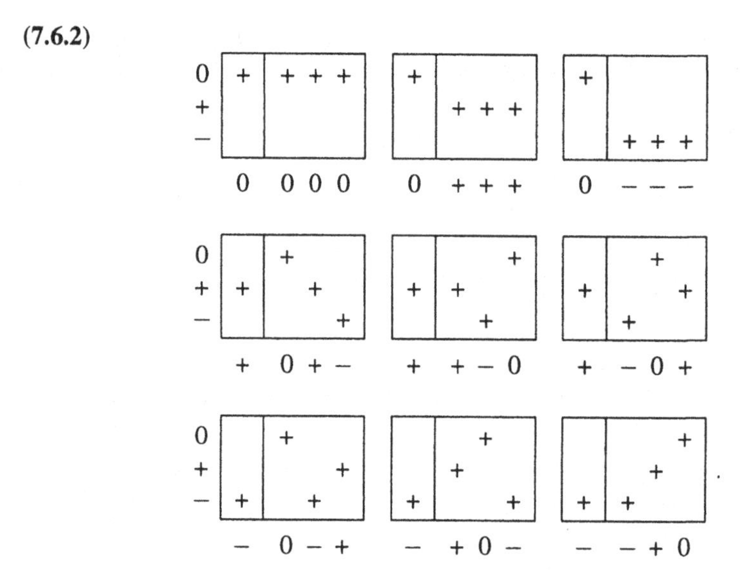

For indices $i, j, \ldots$, define $C_{ij\ldots}$ to be $C_i + C_j + \ldots$.

(7.7) Notation. The *ternary Golay code* $\mathcal{TG}$ is the space of all labled subsets (x_{ci}) of Δ defined by these conditions:

(7.7.1)
$$\sum_c x_{ci} - \sum_c x_{cj} = 0\,, \quad \text{for all } i, j\,;$$

(7.7.2)
$$\sum_c x_{ci} + \sum_c x_{ci}\,((t))_{ci}\,, \quad \text{for all } i \text{ and all } t \in \mathcal{T}\,.$$

Note that (7.7.1) can be interpreted by a dot product with $C_i - C_j$ and that (7.7.2) can be interpreted by a dot product with $C_i - ((t))$. In analogy with Section 5, we call the condition (7.7.1) the *balance condition* and condition (7.7.2) for $t = (0000)$ is called the *well-balanced condition.*

(7.8) Lemma. *The linear space of conditions determined by* (7.7.1) *and* (7.7.2) *corresponds to a 6 dimensional subspace of the dual of* $\mathbb{F}_3^6$. *It has a basis consisting* *of taking the dot products with $C_i - C_j$ (i fixed, $j \neq i$) and $C_i + (t)$, where i is fixed and t ranges over the three vectors of an affinely independent set in $\mathcal{T}$.*

Proof. Clearly, $C_i - C_j$, with i fixed and $j \neq i$, form a basis for the span of all $C_i - C_j$. Let $\mathcal{A}$ be any affinely independent set in $\mathcal{T}$. Define $U := span\{C_i - C_j, C_i + (t) \mid j \neq i, t \in \mathcal{A}\}$.

We claim that U is 6-dimensional and that the spanning set is a basis. Given a dependence relation $\sum_{j \neq i} a_j \left(C_i - C_j\right) + \sum_{t \in \mathcal{A}} b_t (C_i - (t)) = 0$, we apply $\mathcal{L}$ (7.5) to get $\sum_{t \in \mathcal{A}} b_t t = 0$. Taking k-th coefficients, $\sum_{t \in \mathcal{A}} b_t t_k = 0$, for all k. Let $\pi_k : \mathbb{F}_3^{\Delta} \to V_k$ be the projection, where V_k is the subspace supported at K_k. Applying π_j for $j \neq i$, we deduce that $0 = a_j - \sum_{t \in \mathcal{A}^t} t_j = a_j$. Applying π_i, we deduce that $0 = \sum_{t \in \mathcal{A}} b_t + \sum_{t \in \mathcal{A}} b_t t_i = \sum_{t \in \mathcal{A}} b_t$. We now use affine independence of $\mathcal{A}$ to deduce that each $b_t = 0$. The claim follows.

Taking inner products with vectors of the spanning set we deduce that $U \leq U^{\perp}$. We have $U \leq \mathcal{TG}$ since the spanning set for U satisfies (7.7). The condition (7.7) for membership in $\mathcal{TG}$ implies that $\mathcal{TG} \leq U^{\perp}$. Nonsingularity and $dim\, U = 6$ force equality: $U = \mathcal{TG} = U^{\perp}$. □

(7.9) Lemma. (i) *The nonzero vectors*

$$C_i - C_j,\ C_i + (t)\ ,\ (t) - \left(t'\right)\ \text{and}\ C_i + C_j - (t)$$

and their negatives form a set of $12+2\cdot4\cdot9+9\cdot8+2\cdot6\cdot9 = 12+72+72+108 = 264$ *vectors of weight* 6 *in* $\mathcal{TG}$. *Their respective column distributions are* $3^2 0^2$, $3^1 1^3$, $0^1 2^3$ *and* $2^2 1^2$.

(ii) *They span* $\mathcal{TG}$ *and so* $\mathcal{TG}$ *is a self-orthogonal code.*

(iii) *The minimum weight of* $\mathcal{TG}$ *is* 6.

(iv) *If u and v are codewords whose supports are the same* 6*-set, then* $u = \pm v$.

Proof. The cardinality and column distribution statements are obvious and, since $dim\, \mathcal{TG} = 6$ and $264 > 3^5$, so is the spanning statement. Statement (ii) follows if we show that any two of these vectors are orthogonal. The first two kinds actually *span* $\mathcal{TG}$, by (7.6), and it is easy to see that any two such are orthogonal.

We argue that no weight is 3. If $S \in \mathcal{TG}$ has weight 3, then (7.7.1) implies that all column sums for S are 0. Therefore, $supp(S) \cap K_i$ is even, for all i, which is impossible. So, (iii) holds. At once, (iii) implies (iv). □

(7.10) Lemma. (i) *Let A be a k-subset of Δ and define the subspace* $\mathcal{TG}_A := \{w \in \mathcal{TG} \mid supp(w) \cap A = \emptyset\}$. *If $k \leq 5$,* $dim\, \mathcal{TG}_A = 6 - k$. (ii) *The number of weight* 6 *words is* $\binom{12}{5} / \binom{6}{5} \times 2 = 264$.

Proof. (i) The condition that a point be avoided in the support is a linear one, so it suffices to treat the case $k = 5$. Suppose that u and v are nonzero vectors in $\mathcal{TG}_A$. Since $wt(u \pm v)$ is divisible by 3 and is at most 7, it is 0 or 6. We conclude that $supp(u) = supp(v)$ and finally that $u = \pm v$. (ii) Any weight 6 word occurs in exactly 6 subspaces $\mathcal{TG}_A$, where A is a 5-set. □

(7.11) Lemma. *If A is a 3-set, the weight distribution of $\mathcal{TG}_A$ is* $0^1 6^{24} 9^2$.

Proof. To a pair $\pm u$ of weight 6 codewords in $\mathcal{TG}_A$ we associate a triple of 2-sets in $\Delta \backslash (A \cup supp(u))$. By (7.10), three such 2-sets lead to a unique pair of weight 6 words. So, in $\mathcal{TG}_A$ we count $\binom{9}{2}/\binom{3}{2} \times 2 = 24$ weight 6 words. The remaining 2 words must have weight 9. □

(7.12) Lemma. *The weight distribution of $\mathcal{TG}$ is* $0^1 6^{2\times 132} 9^{220} 12^{2\times 12}$.

Proof. By (7.9.ii,iii), all weights are in $\{6, 9, 12\}$. To a pair $\pm w$ of weight 9 words we associate a 3-set (by taking the complement of their support). We use the bijection of (7.11) to see that the number of weight 9 codewords is $2 \cdot \binom{12}{3} = 2 \cdot 12 \cdot 11 \cdot 10/6 = 440$. The remaining 24 codewords must have weight 12. □

(7.13) Definition. A 6-set which is the support of a weight 6 codeword is called a *special hexad* or simply a *hexad*.

(7.14) Lemma. *Let G be any $[12, 6, w]$ ternary code with $w \geq 6$. Then $w = 6$ and every coset may be given a uniquely defined weight $n \in \{0, 1, 2, 3\}$, according to whether the coset contains a word of weight n. The set of elements in a given coset of weight ≤ 3 is a singleton unless the weight is 3 in which case the set consists of four weight 3 vectors. The cocode (3.10) has weight distribution* $0^1 1^{2\times 12} 2^{2\times 132} 3^{2\times 220}$.

Proof. Note that $\binom{12}{3} = 220$. The argument goes as in the proof of (5.9). □

(7.15) Theorem. (i) *The 132 special hexads form a Steiner system $S(5, 6, 12)$ (see (5.1)).*

(ii) *The complement of a special hexad is a special hexad.*

Proof. Let $\mathcal{S}$ be this family of 132 6-sets. We are given a 5-set F and must show that there is a unique $A \in \mathcal{S}$ containing F. If A and B are two such sets, let u and v be associated codewords. Since both contain F, u agrees with v or $-v$ in at least 3 places. Since the minimum weight in $\mathcal{TG}$ is 6, $u = \pm v$, whence $A = B$. Thus, it suffices to prove existence of such $A \in \mathcal{S}$.

Let $f \in \mathbb{F}_3^{12}$ have support F and let w be the weight of the coset $f + \mathcal{TG}$. Clearly, $w \neq 0$ and if $w = 1$, such an A exists. If $w = 2$, let t be a weight 2 word in the coset. Then, as $wt(f - t) \leq 7$, $wt(f - t) = 6$ and so $supp(t) \cap supp(f)$ is a 1-set and the coordinates of f and t here are nonzero and negatives. Thus, $F \subseteq supp(f - t) \in \mathcal{S}$.

Finally, suppose that $w \geq 3$. Then $w = 3$ and we have a coset containing four weight 3 vectors $\{v_1, \ldots, v_4\}$ (7.14). Since $wt(f - v_i)$ is at most 8 and is divisible by 3, there is exactly one index i so that $supp(v_i) \cap F$ is a 2-set; for $j \neq i$, $supp(v_j) \cap F$ is a 1-set. On $supp(v_i) \cap F$, the coordinates of v_i and f are opposite and we have $F \subseteq supp(v_i) \cup F \in \mathcal{S}$.

(ii) Let H be a special hexad and let F be a 5-set in $\Delta \backslash H$. If K is the unique special hexad containing F, K and H must be disjoint or else $H \cap K$ is a 1-set and codewords supported at H and K would have nonzero inner product. □

(7.16) Notation. A *foursome* is a set of four weight vectors which occur in a coset of $\mathcal{TG}$ in $\mathbb{F}_3^{\Delta}$ (7.13). A *quartet* is the set of supports of four vectors in a foursome. The *standard quartet* is $K_1, \ldots, K_4$ and the *standard foursome* is $C_1, \ldots, C_4$ (7.6), (7.7). If Φ is a foursome, Φ^0 denotes the associated quartet. For $v \in \Phi$, $v^0 := supp\,(v) \in \Phi^0$.

(7.17) Corollary. (i) *If S is a four-set in Δ, S is contained in exactly* 4 *special hexads. If H and H' are two such, $H = H'$ or $H \cap H' = S$.*

(ii) *Every weight* 6 *codeword is a vector in* (7.9) *or its negative.*

(iii) *If the 4-set S has column distribution $2^2 0^2$, the special hexads containing S have distributions $3^2 0^2$, $2^3 0^1$, $2^3 0^1$ and $2^2 1^2$. Consequently, these distributions characterize the three types of sums (and their negatives) given in* (7.9). *Note that distribution $3^1 2^1 1^1 0^1$ does not occur.*

(iv) (iii) *holds for the distribution with respect to any quartet.*

Proof. (i) follows from the Steiner system property (7.14). (ii) follows from (7.7). (iii) follows from (ii) and inspection of (7.7).

(iv) Suppose that Φ is a foursome and that (iv) fails for the four set S and the quartet Φ^0. Let $\Phi = \{D_i \mid i = 1,2,3,4\}$ and let $J_i := supp\,(D_i)$, $J_{kl\ldots} := J_k + J_l + \ldots$. Now, let H_i, $i = 1,2,3$, be the three hexads over S but distinct from J_{12}. Since (iv) fails, $H_i \cap J_j$ is a 1-set for all $i = 1,2,3$ and $j = 3,4$. We now proceed to seek a contradiction.

For a special hexad, T, let $w\,(T)$ be one of the two codewords in $\mathcal{TG}$ supported at w. Since, for $i \neq i'$, $w\,(H_i) - w\,(H_{i'})$ has weight at most 8, it has weight exactly 6 and so there is a 2-set in F where the coordinates of $w\,(H_i)$ and $w\,(H_{i'})$ agree; call this set $T\left(i, i'\right) = T\left(i', i\right)$. If there were distinct indices i, i', i'' such that $T\left(i, i'\right) \equiv T\left(i, i''\right) \left(mod\langle F\rangle\right)$, then, for a sign $\pm$, $w\,(H_i) - w\,(H_{i'}) \pm w\,(H_{i''})$ has distribution $2^2 3^2$ hence weight 10, a contradiction. It follows that every 2-set or its complement in F occurs as one of the three $T\left(i, i'\right)$. Choose indices so that $\left|T\left(i, i'\right) \cap J_1\right| = 1$; then one of $\{w\,(H_i), w\,(H_{i'})\}$, say $w\,(H_i)$, is orthogonal to D_1. But then, for any $j \in \{3,4\}$, $D_1 - D_j$ is clearly not orthogonal to $w\,(H_i)$, a contradiction to $\mathcal{TG}$ being self orthogonal. So, (iv) holds. □

(7.18) Definition. The *action of* $\mathbb{F}_3^4 \colon Mon(4,3)$ *on* Δ is the HI-representation (1.21). We get an action on $\mathbb{F}_3^{\Delta}$. The action of the subgroup $\mathcal{T} \colon Aut\,(\mathcal{T})$ preserves $\mathcal{TG}$ since it preserves the set (7.9).

(7.19) Theorem. *In the action of $\mathbb{F}_3^4 \colon Mon\,(4,3)$ on $\mathbb{F}_3^{\Delta}$, the subgroup which preserves $\mathcal{TG}$ is the natural $\mathcal{T} \colon Aut\,(\mathcal{T}) \cong AGL\,(2,3)$.*

Proof. The action of the natural $\mathcal{T} \colon Aut\,(\mathcal{T})$ preserves $\mathcal{TG}$. Let H be the subgroup stabilizing $\mathcal{TG}$. Irreducibility of $Aut\,(\mathcal{T})$ on $\mathcal{T}$ (and so on $\mathbb{F}_3^4/\mathcal{T}$, by duality) implies that $H \cap \mathbb{F}_3^4$ is $\mathcal{T}$ or $\mathbb{F}_3^4$. We eliminate $\mathbb{F}_3^4$ by noting that the image of $((0000)) - ((0+++)) \in \mathcal{TG}$ under $(000+)$ is $((000+)) - ((0++-))$ which is not in $\mathcal{TG}$ since it is not orthogonal to $C_1 + ((0+++))$. Finally, since the stabilizer in $Mon(4,3)$ of $\mathcal{T}$ is $Aut(\mathcal{T})$ (by definition), we finish by quoting (7.3). □

(7.20) Theorem. *The stabilizer in $Aut(\mathcal{TG})$ of the standard quartet is $\mathcal{T} : Aut(\mathcal{T}) \times \langle -1 \rangle$.*

Proof. Let H be the stabilizer and let L be the kernel of the action of H on the standard quartet; then $H/L \cong \Sigma_4$. We show that $|L| = 18$. Notice that L embeds in the direct product of four copies of $2wr\Sigma_3 \cong Mon(3,3)$ and that $\mathcal{T} \leq L$. Let D be the subgroup of L acting diagonally.

We first prove that $C_D(\mathcal{T}) = \langle -1 \rangle$. Let $z \in C_D(\mathcal{T}) \backslash \langle -1 \rangle$. Define V_i to be the 3-dimensional subspace supported by K_i. Since $\mathcal{T}$ is transitive on each K_i, z is scalar on each V_i. Without loss, z is trivial on C_1. But then, since $[C_1 - C_k]z$ must be in $\mathcal{TG}$, for all k, we see that $z = 1$. This proves $C_D(\mathcal{T}) = \langle -1 \rangle$.

Next, let $\mathcal{T} \leq U \in Syl_3(L)$. Since U is elementary abelian, the irreducible action of $Aut(\mathcal{T}) \cong GL(2,3)$ on $\mathcal{T}$ and on $\mathbb{F}_3^4/\mathcal{T}$ implies that $|U| = 3^2$ or 3^4. If 3^4, U contains an element u of order 3 such that u fixes three of the $V_j \leq \mathbb{F}_3^{\Delta}$ pointwise, but not the ith. Such u does not carry any $C_i + (t)$ into $\mathcal{TG}$, for $t \in \mathcal{T}$, a contradiction. So, $U = \mathcal{T}$.

We have that $H/L \cong \Sigma_4$ and so the only thing to show is that $|L : \langle \mathcal{T}, -1 \rangle| = 2$. The Frattini argument shows that $H = LN_H(\mathcal{T})$, so if the statement is false, $Y := C_H(\mathcal{T})$ contains $\mathcal{T} \times \langle -1 \rangle$ with index $2^a \geq 1$. Since $\mathcal{T}$ is transitive on K_i and the $C_{\mathcal{T}}(K_i)$ are distinct, for distinct i, the fact that Y centralizes $\mathcal{T}$ means that, for all i, every element of Y fixes and acts semiregularly on each K_i. Therefore, $Y \leq L$ and a Sylow 2-subgroup of Y lies in D and so is just $\langle -1 \rangle$, whence a $= 0$. □

We now set up (7.23), the ternary analogue of the important labeling result (5.28).

(7.21) Definition. Let Φ be a foursome. For $x \in \pm\widetilde{\Delta}$ (7.5), let x^0 be the element of Δ which supports x. Define the set Δ_Φ of *Φ-labeled points* to be $\{\varepsilon x \mid x \in \widetilde{\Delta}, \varepsilon = \pm$ and there is $v \in \Phi$ whose coefficient at x is $\varepsilon\}$. We regard each vector v of Φ to be a 3-subset of Δ_Φ, namely $\{y \in \Delta_\Phi \mid y^0 \in v^0\}$.

A *labeling* is a function $l : \pm\widetilde{\Delta} \to \mathbb{F}_3$ satisfying $l(-x) = l(x)$ and whose restriction to each part of Φ is a bijection.

(7.22) Definition. Given a foursome Φ and labeling l, define the *4-tuple labeling $\mathcal{L} = \mathcal{L}_\Phi$ with respect to Φ* as follows. Let $x = (x_{ci}) \in \mathbb{F}_3^{\Delta}$ be a labeled set, notated as in (7.5). Write x as a linear combination of the basis Δ_Φ; the coefficient at $p \in \Delta_\Phi$ is $x_{ci}^{\Phi} = \varepsilon_{ci} x_{ci}$, where $\varepsilon_{ci} = \pm$ and $\varepsilon_{ci}^{-1} p$ is the labeled point supported at row c, column i. We define a homomorphism $\mathcal{L}_\Phi : \mathbb{F}_3^{\Delta} \to \mathbb{F}_3^4$ by:

(7.22.1) $$\mathcal{L}_{\Phi}(x)_i = \sum_c c x_{ci}^{\Phi} \, .$$

Now, define *the ternary Golay code* $\mathcal{TG}(\Phi, l)$ just as in (7.5), namely the space of all labled subsets (x_{ci}) of Δ defined by these conditions:

(7.22.2) $$\sum_c x_{ci}^{\Phi} - \sum_c x_{cj}^{\Phi} = 0 \, , \text{ for all } i \neq j \, ;$$

(7.22.3) $$\sum_c x_{ci}^{\Phi} + \sum_c x_{ci}^{\Phi}((t))_{ci}^{\Phi} \, , \text{ for all } i \text{ and all } t \in \mathcal{T} \, ,$$

where $((t))_{ci}^{\Phi}$ denotes coefficient with respect to Δ_{Φ}.

(7.23) Theorem. (i) *Given an ordered foursome Φ, there are* 18 *labelings l such that $\mathcal{TG} = \mathcal{TG}(\Phi, l)$.* (ii) $\mathcal{T} : Z(Aut(\mathcal{T})) \cong 3^2{:}2$ *acts regularly on the set of such labelings. (iii) Consequently, $Aut(\mathcal{TG})$ is transitive on ordered foursomes and has order* $2^7 3^3 5 \cdot 11 = 2|M_{12}|$.

Proof. Given (i), (ii) is trivial and (iii) follows as the analogous result (5.30).

We now prove (i). Let C_i be the i-th member of Φ, $K_i = C_i^0$ and let $\mathcal{X}$ be the set of all $(a, b, \gamma) \in A_1 \times \mathbb{F}_3$, where $A_1 := \{(x, y) \in K_1 \times K_1 \mid x \neq y\}$ is the antidiagonal. Let let $\mathcal{Y}$ be the set of all labelings l such that $\mathcal{TG} = \mathcal{TG}(\Phi, l)$. We know that $|\mathcal{X}| = 18$ but do not know that $\mathcal{Y}$ is nonempty. We fix a point $w \in K_2$ and define $\psi: \mathcal{Y} \to \mathcal{X}$ by $l \mapsto (a, b, \gamma)$, where $l(a) = 0$, $l(b) = +$ and $l(w) = \gamma$.

We first show that ψ is injective. Clearly, we know the labels at all points of the 3-set K_1. Let F be the 4-set $K_{12} \setminus \{a, w\}$. Since we have a Steiner system, given any point $p \in \Delta \setminus F$, there is a unique special hexad containing $F \cup \{p\}$. Any two distinct hexads containing F intersect in F (7.17.i). By (7.17.iv), there is a unique such hexad with distribution $2^2 1^2$. It is labeled by $(0\gamma\gamma\gamma)$ and has one point in each of K_3 and K_4; these points are where the label γ goes in K_3 and K_4. So far, we know that 6 values of l are determined.

Next, we consider the 4-set $E := K_{12} \setminus \{b, w\}$. If H is a special hexad containing E other than K_{12}, a hexacode word supported at H is labeled by a hexacode word of slope 1 (7.4) and so its coordinates at columns 2, 3, 4 are distinct. Now let $j \in \{3, 4\}$ and let $p \in K_j$ be labeled by γ. By expanding $K_{1j} \setminus \{b, p\}$ to the hexad H_j with distribution $2^2 1^2$ (7.17.iv), we see that each K_i, for $i \neq 1, j$, gets one point of H_j, and these points are labeled by $\mathbb{F}_3 \setminus \{\gamma\}$ in some order. For $j = 3$ and 4, we deduce labels on disjoint sets in K_{234}. Thus, l is determined.

We next show that ψ is onto. Suppose that (a, b, γ) is given; recall that $w \in K_2$ is fixed. Then, expansion of $K_{12} \setminus \{a, w\}$ and $K_{12} \setminus \{b, w\}$ to special hexads leads to a labeling l of Δ by use of (7.4.i). We must show that $\mathcal{TG}(\Phi, l) = \mathcal{TG}$; see (7.22). Notice that $\mathcal{TG}(\Phi, l)$ contains each $C_i - C_j$ and $C_i + (t)$, for t labeled by $(0\gamma\gamma\gamma)$ and $(+\gamma\gamma'\gamma'')$. Such elements *span* $\mathcal{TG}$. □

(7.24) Corollary. *Up to equivalence of codes, the ternary Golay code is the unique code with parameters* $[12, 6, w]$, *for* $w \geq 6$.

Proof. (7.13) and (7.23). □

(7.25) Proposition. *$Aut(\mathcal{TG})/\langle -1\rangle$ is a simple group which is (sharply) 5-fold transitive on Δ.*

Proof. Imitate the corresponding arguments for M_{24} and M_{12}. It helps to first check that a Sylow 11-normalizer has shape $11:5$. □

(7.26) Notation *(The Standard Basis of $\mathcal{TG}$ and test for membership in $Aut(\mathcal{TG})$).* The labeled sets

(7.26.1)

	+	−	
	+	−	
	+	−	

	+		−
	+		−
	+		−

	+	+	+
	−	−	−

+	−	+	+
	+		
	+		

	−		
+	+	+	
	+		+

	−		
	+		+
+	+	+	

form a basis for $\mathcal{TG}$, *the standard basis*. Since $\mathcal{TG}$ is self orthogonal, a matrix g in $Mon(12,3)$ may be tested for membership in $Aut(\mathcal{TG})$ by checking that it passes the tests $\left(b_i, b_j^g\right) = 0$, for all $i, j = 1, \dots 6$; this is easy by hand and even easier with a computer program. Compare with the analogous (5.35), (5.37).

(7.27) Notation. Let Ω be the 24-set of Sections 5 and 6. The *shuffle labeling* of Δ is the bijection $\Delta \to \{0, 1, 2, \dots, 11\}$ depicted below. The *left interpreter* is the injection $\lambda: \Delta \to \Omega$, defined by

(7.27.1)

6	3	0	9
5	2	7	10
4	1	8	11

$\longrightarrow$

	0		7		8
2		5		10	
1		6		9	
3		4		11	

The *right interpreter* is the injection $\rho: \Delta \to \Omega$ defined by

(7.27.2)

6	3	0	9
5	2	7	10
4	1	8	11

$\longrightarrow$

0		1		2	
	3		4		5
	6		7		8
	9		10		11

Let L and R be the left and right dodecads of (6.20). Then, $Im(\lambda) = L$ and $Im(\rho) = R$. Our definitions of the interpreters are inspired by the so called "shuffle labeling" of Ω [Con-Sl], p. 320.

(7.27.3)

0+	0−	1+	7−	2+	8−
2−	3+	5−	4+	10−	5+
1−	6+	6−	7+	9−	8+
3−	9+	4−	10+	11−	11+

(+ = left, − = right).

Let λ_* and ρ_* be the maps induced by λ and ρ, respectively, of $Mon(12,3)$ to Σ_L, Σ_R, respectively.

(7.28) Notation *(Monomial matrices)***.** Permutation matrices may be notated by the usual cycle depiction of the associated permutation. Let m be a monomial matrix and suppose it has decomposition $m = pd$, where p is a permutation matrix and d is diagonal. If p is a product of cycles $(a, b, \ldots)\ldots$ and d has entry $d_a, d_b, \ldots$ at index $a, b, \ldots$, we denote m by $(d_a a, d_b b, \ldots)\ldots$. We write $\bar{a}$ if $d_a = -1$. For example, $(2\bar{1}\bar{3}) \in Mon(3, F)$ denotes $\begin{pmatrix} 0 & 1 & 0 \\ -1 & 0 & 0 \\ 0 & 0 & -1 \end{pmatrix}$. If the cycles for $p \in \Sigma_X$ are indicated by closed circuits of directed arrows drawn between the points of X, we notate m by drawing the nonzero scalar entries d_a on the arrow which ends at a. We indicate $d_a = +1, -1$ by making the arrowhead $\rightarrow, \twoheadrightarrow$. Thus, $\begin{pmatrix} 0 & -1 \\ 1 & 0 \end{pmatrix}$ may be compactly notated by $1 \longleftrightarrow\!\!\!\!\!\twoheadrightarrow 2$.

(7.29) Proposition. (i) *$Im(\lambda_*)$ gives the image the action of M_{12} on the left dodecad of* (6.20). *In fact, the monomial transformations*

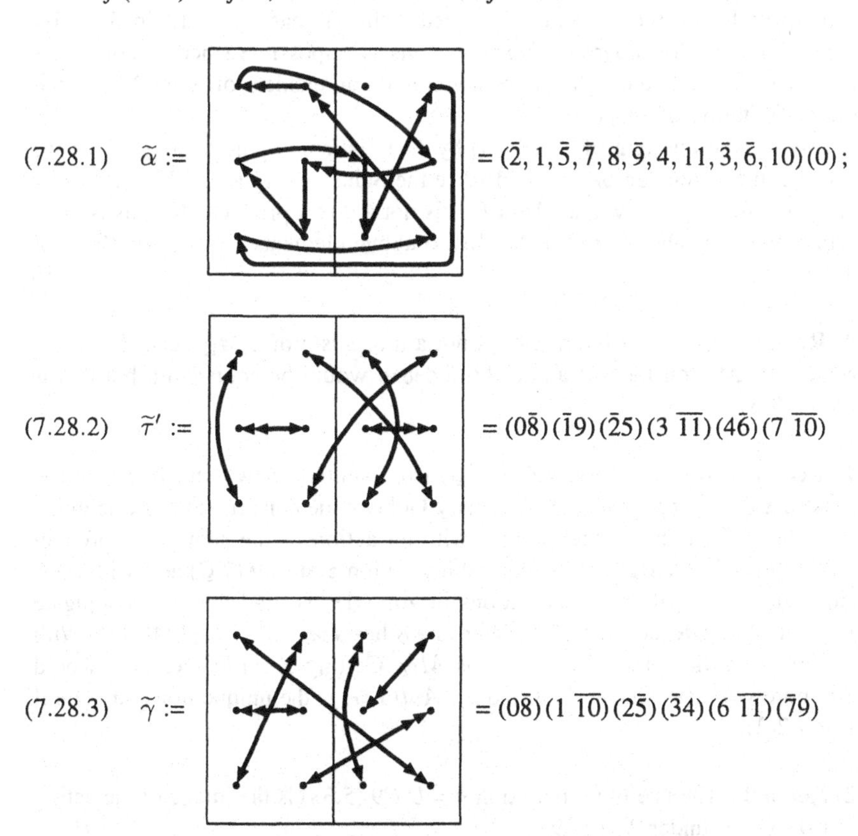

generate $Aut\,(\mathcal{TG})$ *and under* λ_* *go to* α_L, γ_L *and* τ'_L. *Also,* β (6.21) *lifts to*

(7.28.4) $\widetilde{\beta} :=$ 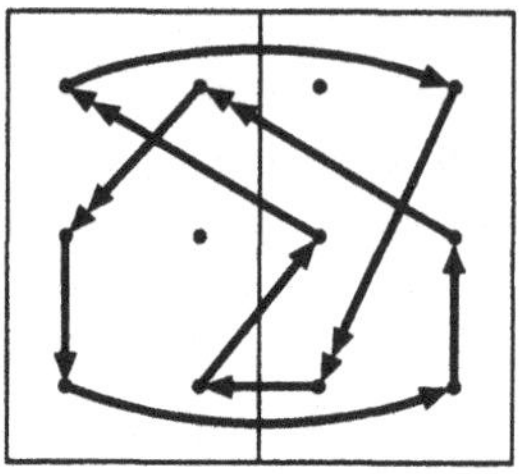 $= (0)\,(2)\,(\bar{6}\,9\,\bar{8}\,1\,7)\,(\bar{3}\,\bar{5}\,4\;11\;10)$;

(ii) $Aut\,(\mathcal{TG}) \cong 2{\cdot}M_{12}$, *the covering group of* M_{12}.

Proof. (i) Membership of each of these in $Aut\,(\mathcal{TG})$ is straightforward with (7.26). Their images under λ_* are straightforward to identify with (6.21) or (5.39). Let $Y := \langle \alpha_L, \gamma_L, \tau'_L \rangle \leq M_{12}^L$, the image of M_{12} in Σ_L, L the left dodecad. To prove generation, we first observe that $\langle \alpha_L, \gamma_L \rangle \cong PSL\,(2,11)$ (6.22.i). Since $\langle \tau', \gamma, \beta \rangle = \langle \tau' \rangle \times \langle \gamma, \beta \rangle \cong 2 \times Dih_{10}$ ((6.21) and (6.22.iii)) and $\beta \in \langle \alpha, \gamma \rangle\,(6.22.i)\,, Y > \langle \alpha_L, \gamma_L \rangle$. If $Y < M_{12}$, the fact that a Sylow 11-normalizer has shape 11:5 in Alt_{12} means that Y has index 12 in M_{12}. By (6.25.3), Y is an M_{11}-subgroup. However, this is impossible since τ' commutes with $\beta \in \langle \alpha, \gamma \rangle$ (6.21) and Alt_{11} does not contain an element of order 10 centralizing a permutation of shape $1^2 5^2$.

(ii) By (i), the quotient of $Aut\,(\mathcal{TG})$ by $\langle -1 \rangle$ is isomorphic to M_{12}. Since we have a a central extension of M_{12} and since the Schur multiplier of M_{12} has order 2 (2.14), it suffices to show that $Aut\,(\mathcal{TG})$ is not a direct product. But this is clear since involutions γ and τ' in $Im\,(\lambda_*)$ lift to above elements $\widetilde{\gamma}, \widetilde{\tau}' \in Aut\,(\mathcal{TG})$ of order 4 (2.28). □

(7.30) Remark. The right interpreter is not a morphism of $2{\cdot}M_{12}$ sets. If it were, the actions of M_{12} on the left and right dodecad would be equivalent, but this is not so (6.18.ii).

(7.31) Remark. We can analyze $Aut\,(\mathcal{TG})$ from another viewpoint. In Section 9, we construct the Leech lattice, Λ. A ternary Golay code construction can be given in the style of Section 9 which comes with an action of the semidirect product $\mathcal{TG} : Aut\,(\mathcal{TG})$. It is clear that the central involution z of $Aut\,(\mathcal{TG})$ acts with trace 0. Using the classification of involutions in $Aut\,(\Lambda)$, we find that z is conjugate to $\varepsilon_{\mathcal{D}}$, with $\mathcal{D}$ a dodecad. The centralizer of $\varepsilon_{\mathcal{D}}$ has shape $2^{12} : M_{12}$ (10.15). With (7.25), this identifies $Aut\,(\mathcal{TG})/\langle -1 \rangle$ as M_{12}. Getting $Aut\,(\mathcal{TG})$ exactly would require more work (as we saw in (7.28.ii), $Aut(\mathcal{TG})$ is the unique nonsplit central extension $2{\cdot}M_{12}$).

(7.32) Remark. The useful permutation $\mu = UP9$ (5.38) is the image of the action of $(-+0-) \in \mathcal{T}$ under λ_* (7.29).

(7.33) Exercise. $AGL(2,3)$ has three classes of elements of order 3. Hint: Besides the nonidentity elements in the group of translations, T, we have the cosets of the form Tx, where $x \in GL(2,3)$ has order 3 (there is one $GL(2,3)$-class of such). The elements of Tx all have order 3 and fall into two orbits under the action of $TC_{AGL(2,3)}(x) \cong 3^2{:}[3 \times 2]$.

(7.34) Exercise. (i) Prove that a solvable doubly transitive group on 9 points has a regular normal subgroup and that it is complemented by a group isomorphic to $Quat_8$, SD_{16} or $GL(2,3)$.

(ii) Prove that any sharply 5-transitive group G on Δ has the property that the stabilizer of a 3-set T is isomorphic to $AGL(2,3)$, induces Σ_3 on T and acts faithfully on $\Delta \backslash T$. Conclude that such G has exactly two conjugacy classes of elements of order 3 and that their cycle shapes are $1^3 3^3$ and 3^4. Hint: (i) and (7.33).

(7.35) Remark. Once we know that M_{12} has two classes of elements of order 3, it follows that M_{24} does too since M_{24} contains elements with cycle structures $1^6 3^6$ (e.g. UP9) and 3^8 (e.g. look in the sextet group or the trio group; or see UP19).

(7.36) Exercise. Verify that there are four orbits of $Aut(\mathcal{TG})$ on $\mathcal{TG}$, according to weights.

(7.37) Exercise. Determine that the Jordan canonical forms for elements of order 3 in $2\cdot M_{12}$ on $\mathcal{TG}$ are

(7.37.1) $2J_3$ if the element of order 3 has cycle shape 3^4;

(7.37.2) $J_1 J_2 J_3$ if the element of order 3 has cycle shape $1^3 3^3$.

Chapter 8. Lattices

We develop enough lattice theory to analyze the Leech lattice and several other lattices of interest.

(8.1) Definition. A *lattice* is a free abelian group L of finite rank with a symmetric bilinear form $f : L \times L \to A$, where A is an abelian group and $Im(f)$ is 0 or infinite cyclic. We call L an *integral lattice* in case $Im(f) \subseteq \mathbb{Z}$. If K is a characteristic 0 field, $K \otimes_{\mathbb{Z}} L$ is the *ambient K-vector space.*

(8.2) Remark. Usually, A is an additive subgroup of $\mathbb{C}$ of the form $q\mathbb{Z}$, for some complex number q. In this section, q will be rational. The definition (8.1) may seem a little cumbersome to those who think that integer lattices are the essential case. However, it is convenient to use the term lattice for free abelian groups containing an integer lattice with finite index and certain other cases, for instance, the dual of an integer lattice. The exercise below indicates some properties of lattices as subgroups of Euclidean space without reference to the values of the inner product.

(8.3) Exercise. Let L be a finitely generated subgroup of the additive group of $V \cong \mathbb{R}^n$. Show that the following are equivalent: (a) L is free abelian of rank n and contains a basis of V; (b) L is free abelian of finite rank and closed in V and V/L is a compact group (hence a product of n circles, $\mathbb{R}/\mathbb{Z}$).

In case there is an inner product in (8.3), (L, L) may be a noncyclic (even dense) subgroup of $(\mathbb{R}, +)$, e.g., $L = span\,\{(1,0), (\pi, 1)\}$, with the usual dot product.

(8.4) Definition. The lattice is *nondegenerate* if $x \in L$ and $f(x, L) = 0$ implies that $x = 0$. The *radical* of L (or, more precisely, of f) is $\{x \in L \mid f(x, L) = 0\}$.

We often use parentheses (.,.) for the symmetric bilinear form, understood from context.

(8.5) Definition. Let L be a lattice and $x_1, \ldots, x_n$ a basis. The *Gram matrix* is the $n \times n$ matrix $A = ((x_i, x_j))$. Let $y_1, \ldots, y_n$ be another basis, $B = ((y_i, y_j))$. Let $A = (a_{ij})$, $B = (b_{ij})$ and define $C = (c_{ij})$ by $y_i = \sum c_{ij} x_j$. Then $b_{ij} = \sum_{k,l} c_{ik} a_{kl} c_{jl}$, whence $B = CAC^t$. Since $det(C) = \pm 1$, $det\, A = det\, B$, and so this scalar is an invariant of L, called the *determinant* of L. In the positive definite case, it may be interpreted as the square of the volume of the parallelopiped spanned by the basis vectors.

The dimension of the radical is the nullity of the Gram matrix and the form is nondegenerate iff the Gram matrix is nonsingular.

(8.6) Definition. If L is a lattice, $Aut(L)$ is the group of automorphisms of L as an abelian group which preserve the bilinear form.

(8.7) Remark. $Aut(L)$ preserves all balls about the origin. Assume L is positive definite. The intersection of any compact set (e.g., a ball) with L is finite. The intersection of a sufficiently large ball with L spans the ambient real vector space, whence $Aut(L)$ is finite. In case L is not positive definite, $Aut(L)$ may be finite or infinite. For instance, if the radical is nonzero, $Aut(L)$ is infinite. On the other hand, if L has Gram matrix $\begin{pmatrix} 0 & 1 \\ 1 & 0 \end{pmatrix}$, L is indefinite but $Aut(L) \cong \mathbb{Z}_2 \times \mathbb{Z}_2$, a finite group.

(8.8) Remark. The above definitions are extensions of the usual concepts for bilinear forms on vector spaces. Let L and q be as in (8.2). If K is any field of characteristic 0 and p some fixed element of $K^\times$, a lattice gives rise to a bilinear form $F : V \times V \to K$, where $V = K \otimes_{\mathbb{Z}} L$, $F(c \otimes x, d \otimes y) = cdep$, and $f(x, y) = eq$. This form F is characterized by commutatitivity of the following diagram

$$\begin{array}{ccccc} f : L \times L & \longrightarrow & \mathbb{Z}_q & & q \\ \downarrow & & \downarrow & & \downarrow \\ F : V \times V & \longrightarrow & \mathbb{Z}_p\,, & \text{where} & p \end{array}$$

and where the left vertical map is the natural one. In case $Im(f) \leq K$, we take $p = q$ and the definition reads $F(c \otimes x, d \otimes y) = cdf(x, y)$.

(8.9). Definition. The *signature* of a lattice L with real-valued bilinear form f is the signature of the form f extended to the ambient real vector space, as in (8.8) (with $p = q$). So, we may speak of a *positive definite lattice* (signature $(n, 0, 0)$), a *hyperbolic lattice* (signature $(n-1, 1, 0)$), *indefinite lattice* (signature (p, q, r), with $r > 0$), etc.

(8.10) Definition. We assume that the lattice L is embedded in a K-vector space V ($char\, K = 0$) as above (so that L spans V and $rank\, L = dim\, V$). The *dual* of L is $L^* = \{x \in V \mid F(x, L) \subseteq \mathbb{Z}\}$.

As indicated in (8.8), if p is not specified and $q \in \mathbb{Q}$, we take $p = q \in \mathbb{Q} \leq K$. It follows that $L^* \leq \mathbb{Q}L$, so L^* is in a sense independent of K.

If $det(L) \neq 0$, $(L^*)^* = L$ (Hint: a basis dual to a basis for L is a basis for L^*).

(8.11) Examples. *The A_1-lattice.* $L = \mathbb{Z}\alpha$, where $f(\alpha, \alpha) = 2$. Then $L^* = \frac{1}{2}\mathbb{Z}\alpha$.

The A_2-lattice. L has basis α, β and bilinear function f satisfying $f(\alpha, \alpha) = 2$, $f(\beta, \beta) = 2$, $f(\alpha, \beta) = -1$. Its determinant is 3. The dual lattice is just the $\mathbb{Z}$-span of $\frac{1}{3}(\alpha - \beta)$ and either of α or β. The dual lattice is isometric to $\frac{1}{\sqrt{3}} L$.

In each of the above cases, f has an extension to F on L^*; $Im(f) = \mathbb{Z}$ and $Im(f) = \frac{1}{2}\mathbb{Z}$, $\frac{1}{3}\mathbb{Z}$, respectively. Thus, in each case, L is an integral lattice while L^* is not.

(8.12) Lemma. *Let L be a lattice with sublattice M of finite index. Then, $\det M = |L : M|^2 \cdot \det L$. The same formula holds if L and M are lattices in the same rational vector space with $L \cap M$ of finite index in both and if $|L : M|$ is interpreted to mean $|L : L \cap M| / |M : L \cap M|$.*

Proof. It suffices to prove the first statement. The theory of modules over a principal ideal domain implies that there are a basis $x_1, \dots, x_n$ of L and integers $d_1, \dots, d_n$ such that $d_1 x_1, \dots, d_n x_n$ is a basis of M. A comparison of Gram matrices immediately implies the formula. □

(8.13) Lemma. *Let L be an integral lattice and let $\varphi : L \to Hom(L, \mathbb{Z})$ be the natural map defined by $\varphi(x)(y) = (x, y)$. Then, $Im(\varphi)$ has index $|\det(L)|$. Moreover, the cokernel $(\cong L^*/L)$ is isomorphic to the direct sum of cyclic groups $\mathbb{Z}/(d_i)$, where $d_1, \dots, d_n$ is the set of elementary divisors of a Gram matrix for L.*

Proof. Standard matrix theory from the theory of modules over a principal ideal domain (e.g., [Jac]) tells us that there are a basis $f_1, \dots, f_n$ of $Hom(L, \mathbb{Z}) \cong \mathbb{Z}^n$ and scalars $d_1, \dots, d_n$ with $d_i | d_{i+1}$ for $i = 1, \dots, n$ such that $d_1 f_1, \dots, d_n f_n$ is a basis for $Im(\varphi)$. We are done once we show that the d_i are the elementary divisors of a Gram matrix for L. In particular, if $\det(L) \neq 0$, $|L^*/L| = |\det(L)|$.

Let $e_1, \dots, e_n$ be the basis of $\mathbb{Q} \otimes L$ dual to $f_1, \dots, f_n$. Then $e_1, \dots, e_n$ is a basis for L. Define the matrix $Q = (q_{ij})$ by $e_i = \sum q_{ij} v_j$; then $\det Q = \pm 1$. Let $v_i = \varphi^{-1}(d_i f_i)$, $i = 1, \dots, n$, another basis for L. Then $(v_i, v_j) = d_i f_i(v_j)$, and the matrix $(f_i(v_j))$ may be transformed to the $n \times n$ identity matrix by right multiplication with ${}^tQ$. It follows that $((v_i, v_j))^t Q = diag(d_1, \dots, d_n)$. □

(8.14) Definition. If (L, f) is a lattice and $Im(f) = \mathbb{Z}q$, the *even sublattice* is $\{x \in L | f(x, x) \in 2\mathbb{Z}q\}$. It has index 1 or 2 and is the kernel of the natural homomorphism $L \to \mathbb{Z}/2\mathbb{Z}$ given by $x \mapsto f(x, x)/q \pmod 2$ when $q \neq 0$. A lattice is *even* if it coincides with its even sublattice. A *root* is a lattice vector x such that $f(x, x) = 2$.

(8.15) Remark. An integral lattice is even iff for any spanning set S, $f(x, y) \in \mathbb{Z}$ and $f(x, x) \in 2\mathbb{Z}$ for all $x, y \in S$.

(8.16) Proposition. *Let L be a nonsingular integral lattice. Let M be a subgroup of L which is a direct summand of L as abelian groups. We study the annihilator of M in L and L^*.*

(a) *If L is unimodular, $det\, M = det\, ann_L(M)$.*

(b) *Suppose $det\, M \neq 0$. The natural map $L \to Hom(M, \mathbb{Z})$ is a subgroup of finite index dividing $|det\, L|$; the cokernel is isomorphic to a subgroup of L^*/L and has order $|det\, M|$; $det\, ann_L(M) = (e^2/d)\, det\, M$, where $d = det\, L$ and e is a divisor of d.*

Proof. (a) is trivial if $det\, M = 0$ and it follows from (b) if $det\, M \neq 0$ by taking $d = 1$, so we prove (b). Let $x_1, \ldots, x_r$ be a basis for M and extend it to a basis $x_1, \ldots, x_n$ for L, $n \geq r$. Let $y_1, \ldots, y_n$ be the dual basis (in the ambient rational vector space).

Now let $\varphi : L^* \to M^*$ be the natural epimorphism and let $\pi : M^* \to M^*/M$ be the natural quotient. Set $\psi = \varphi\pi$; $Ker\,(\psi) = ann_{L^*}(M) + M$. Set $\mu = det\, M \neq 0$, $d = det\, L$. Note that $ann_{L^*}(M) = span\,\{y_{r+1}, \ldots, y_n\}$ and that, by (8.13), M has index $|\mu|$ in $span\{y_1, \ldots, y_r\}$. By (8.12), $det(Ker\, \psi) = \mu^2\, det\,(L^*) = \mu^2 d^{-2} d^1 = \mu^2 d^{-1}$.

Since M is nonsingular, $A := (ann_{L^*}(M) + M)\,/\,(ann_L(M) + M) \cong ann_{L^*}(M)/\, ann_L(M) = ann_{L^*}(M)/L \cap ann_{L^*}(M) \cong ann_{L^*}(M) + L/L$, which is embedded in L^*/L. It follows that A is an abelian group of order e, for some divisor e of d. Therefore, by (8.12), $det\,(ann_L(M) \perp M) = e^2\, det\,(Ker\, \psi) = \mu^2 e^2/d$ and, since the sum is orthogonal, $det(ann_L(M) \perp M) = \mu\, det(ann_L(M))$, and the result follows. □

(8.17) Exercise. Study the situation for the lattice $L = L_1 \perp L_2$, where L_1 and L_2 are given by respective Gram matrices (2) and $\begin{pmatrix} 2 & -1 \\ -1 & 2 \end{pmatrix}$ (so L_i is the A_2 root lattice; see (8.11)). Compute e as in (8.16) for each sublattice $M = L_1$, L_2 and a sublattice of L_2 with Gram matrix (2).

(8.18) Examples. We describe several standard lattices, all of which are integral and positive definite. Note that, in the case of root lattices, (8.16) gets the determinant of the Cartan matrix more easily than a direct calculation.

The *square lattice* SQ_n has $\mathbb{Z}$-basis $x_1, \ldots, x_n$ with $(x_i, x_j) = \delta_{ij}$. Its determinant is 1.

The *A_n-lattice* L_{A_n} is the subgroup of SQ_{n+1} consisting of all $\sum c_i x_i$ with $\sum c_i = 0$. It is the annihilator of the sublattice spanned by $\nu := \sum x_i$. Since $(\nu, \nu) = n + 1$, the A_n-lattice has determinant $n + 1$, by (8.16).

The *D_n lattice* L_{D_n} is the even sublattice of the square lattice of rank n. From (8.16), its determinant is 4. If $L = L_{D_n}$, the quotient L^*/L is $\mathbb{Z}_4$ or $\mathbb{Z}_2 \times \mathbb{Z}_2$ as n is odd or even; it is generated by the images of $\nu/2$ and $\{\nu/2, x_1\}$ as n is odd or even, respectively.

The *E_8-lattice* L_{E_8} is $L_{D_8} + \mathbb{Z}\,[\nu/2]$. By (8.16), its determinant is 1. One may define an E_7- and E_6-sublattice as the annihilators of $\mathbb{Z}\alpha$ and $\mathbb{Z}\alpha + \mathbb{Z}\beta$, respectively, where $(\alpha, \alpha) = (\beta, \beta) = 2$ and $(\alpha, \beta) = -1$. Such vectors α and β exist and the automorphism group of the E_8-lattice operates transitively on such configurations, so the definition is unambiguous. We do not prove this statement here but refer to [Bour] or [Hum] for properties of root systems.

An important fact, proved by E. Witt, is that the E_8-lattice is, up to isometry, the unique even integral unimodular lattice of rank 8 [Serre] [MH].

(8.18) Example. The *halfspin lattice of rank* n *is* $HS_n := L_{D_n} + \mathbb{Z}(\nu/2)$ has determinant $\frac{1}{4}$ or 1 as n is odd or even, is integral iff $n \equiv 0\,(mod\,4)$ and is even integral iff $n \equiv 0(mod\,8)$. It is isometric to $L_{D_n} + \mathbb{Z}(\nu_i'/2)$, where $\nu_i' = \nu - 2x_i$ (apply the reflection at x_i to see this). The set of roots in it spans a lattice isometric to $HS_8 \cong L_{E_8}$ if $n = 8$ and L_{D_n} if $n \neq 8$.

(8.19) Example. Let L_0 be a unimodular even integral lattice and set $L := \sqrt{2} \cdot L_0$. Then $L^* = \frac{1}{2} \cdot L$. The space L^*/L has an $\mathbb{F}_2$-valued bilinear form Q on it, namely $x+L \mapsto (x,x)\,(mod\,2)$. If J/L is a subspace of L^*/L, J/L is isotropic with respect to Q iff J is an even integral lattice and J is unimodular iff $|J/L|^2 = |L^* : L|$.

(8.20) Definition. Let L be an even integral lattice. The set $L_n := \{x \in L \mid (x,x) = 2n\}$ is called the set of vectors of *type* n. In the positive definite case, the *theta function* of L is the series $\sum_{n=0}^{\infty} a_n q^n$, where $a_n := |L_n|$. The symbol q is interpreted as a formal variable or as the complex variable $e^{2\pi i z}$, where $z \in \{w \in \mathbb{C} \mid Im(w) > 0\}$.

A proper treatment of theta functions requires the theory of modular forms, which we shall not attempt to summarize here. For example, see references [MH] [Ogg] [Serre] [Shi]. One of the few properties which we use in this book is that the series is determined if we know a few initial coefficients.

(8.21) A few theta functions.

$$E_8\text{-lattice: } 1 + 240q + 2160q^2 + 6720q^3 + 17520q^4 + \dots\,.$$

$$HS_{16}\text{: } 1 + 480q + 61920q^2 + 1050240q^3 + \dots\,.$$

(This is the same as the square of ϑ_{E_8}-series and so the same as the theta series for the orthogonal direct sum of two E_8-lattices; thus, the theta series does not generally determine the lattice.)

(8.22) Exercise *(Alternate constructions of the* E_8*-lattice).* The basic description of the Leech lattice (see (9.1) and (9.2)) has an analogue for the lower rank E_8-lattice.

(8.22.1) Let H be the $[8,4,4]$ extended Hamming code on the alphabet X; see (3.14.b). Let L_0 be the rank 8 square lattice with orthonormal basis $\{e_i \mid i \in X\}$. Let $L := \sqrt{2} \cdot L_0$, $f_i := \sqrt{2} \cdot e_i$, $M := span\{L, \frac{1}{2}\sum_{i\in A} f_i \mid A \in H\}$. Then, by (8.19), M is an even, integral unimodular lattice of rank 8, whence, by (8.17), M is isometric to L_{E_8}. Conclude that $Aut(M)$ contains a group of the form $2^{8+3}.\,GL(3,2)$.

(8.22.2) Use $\mathcal{T}$, the tetracode (7.1) to construct the E_8-lattice by carrying out the following steps.

(i) Let $M = M_1 \perp \ldots \perp M_4$ be a sublattice of $\mathbb{R}^8$, where each M_i is isometric to the A_2-lattice. Thus, each M_i has a basis of the form α_i, β_i which is a system of fundamental roots of the root system in M_i.

(ii) Let M_i^* be the dual lattice of M_i in the rational vector space $\mathbb{Q} \otimes M_i$. Let $\nu_i = \frac{1}{3}(\alpha_i - \beta_i)$. Then $(\nu_i, \alpha_i) = 1, (\nu_i, \beta_i) = -1, (\nu_i, \nu_i) = \frac{2}{3}$ and $M_i^* = M_i + \mathbb{Z}\nu_i$.

(iii) Let $I = \{1, 2, 3, 4\}$ and consider a standard tetracode $\mathcal{T}$ indexed by I. For a codeword $c = (c_i)$, define $\nu_c := \sum_{i \in I} \widehat{c_i} \nu_i$, where $\widehat{a}$ denotes the integer in $\{-1, 0, 1\}$ which reduces modulo 3 to $a \in \mathbb{F}_3$. Define $L := M + \sum_{c \in \mathcal{T}} \mathbb{Z}\nu_c$.

(iv) Prove that L is an even, unimodular lattice, hence is isometric to the E_8-lattice.

(8.22.3) We continue (8.22.2) to get information about automorphisms of the E_8-lattice.

(i) Let J_i be the Weyl group of the root system spanned by α_i and β_i, considered as a group of linear transformations on the ambient vector space of M. Then, $J_i \cong \Sigma_3$. Prove that $Aut(M_i) = J_i \times \langle z_i \rangle$, for all i, where z_i is -1 on M_i and is 1 on M_j, for $j \neq i$. Thus, $Aut(M_i) \cong \Sigma_3 \times 2 \cong Dih_{12}$.

(ii) There is a natural $Dih_{12}\, wr \Sigma_4$ acting on M, suggested by the decomposition of (8.22.2). The intersection of this group with $Aut(L)$ is a group $[J_1 \times \ldots \times J_4] : Aut(\mathcal{T}) \cong \Sigma_3 wr\, GL(2,3)$.

(8.22.4) The preceeding constructions have analogues for any rank 8 subdiagram Γ of the extended Dynkin diagram for E_8. In the other cases, the group we get is relatively small augmentation of the Weyl group for Γ. For instance, if Γ has type A_4^2, the Weyl group is $\Sigma_5 \times \Sigma_5$ and the subgroup of the Weyl group of E_8 we are led to is just $\Sigma_5 wr 2 \times \langle -1 \rangle$ (the wreathing "2" is due to an involution interchanging the two A_4-sublattices). In any case, this style of construction displays many subgroups of the Weyl group of E_8.

(8.23) Classifications of even integral unimodular positive definite lattices. The dimension must be divisible by 8.

Dimension 8: the E_8-lattice. [Witt37].

Dimension 16: HS_{16} and $L_{E_8} \perp L_{E_8}$. [Witt37].

Dimension 24: there are 24 isometry types of such lattices, as classified by Niemeier [Nie67]. They are distinguished by the systems formed by the roots in the lattices. The root systems which occur are listed below; accordingly, we write $\mathcal{N}_\Phi$ for the unique such lattice with root system of type Φ and call it *the Niemeier lattice of type* Φ. A short proof was published by Venkov [Ven78], who makes use of modular forms and uniqueness of error correcting codes of small length over finite rings to deduce the classification.

Root systems for Niemeier lattices of rank 24

$\emptyset$;

A_1^{24}, A_2^{12}, A_3^8, A_4^6, A_6^4, A_8^3, A_{12}^2, A_{24};

D_4^6, D_6^4, D_8^3, D_{12}^2, D_{24};

E_6^4, E_8^3;

$A_5^4D_4$, $A_7^2D_5^2$, $A_9^2D_6$, $A_{15}D_9$, $D_{16}E_8$, $D_{10}E_7^2$, $A_{17}E_7$;

$A_{11}D_7E_6$.

Rank at least 32: there are at least 10^7 isometry types.

(8.24) Exercise. Let L be the E_8-lattice and let R be the set of 240 roots.

(8.24.1) Show that two elements of R are congruent modulo $2L$ iff they are equal or opposite;

(8.24.2) Show that $F := \{x + y \mid x, y \in R, (x, y) = 0\}$ consists of $2160 = 2^4 \cdot 3^3 \cdot 5$ vectors of norm 4;

(8.24.3) For $u \in F$, $\{v \in F \mid u - v \in 2L\}$ is a *frame*, i.e., a set of 16 elements of norm 4, two of which are equal, opposite, or orthogonal;

(8.24.4) A nontrivial coset of $2L$ in L meets $R \cup F$ in either $\{x, -x\} \subseteq R$ or a frame;

(8.24.5) Conclude that F is the full set of vectors of norm 4 in L.

The above is an analogue of our results on the Golay cocode and the Leech lattice modulo 2. Knowledge of the theta function of L and (8.24.2) would prove (8.24.1).

(8.25) Exercise. Show that the reduction modulo 2 of the lattices of types E_6, E_7 and E_8 gives an isomorphism of the respective Weyl groups onto an orthogonal group over $\mathbb{F}_2$. Hint: (a) show that the reflections induce orthogonal tranvections and the right number of those; (b) show that the kernel of these maps are central and of orders $1, 2$ and 2; (c) show that the central extension is split in every case but E_8.

Chapter 9. The Leech Lattice and Conway Groups

We describe the Leech lattice and obtain some basic information about its automorphism group. Our presentation is based mainly on Sections 5 and 8.

(9.1) Notation. We use the following notation:

$(-1)^X := (-1)^{|X|}$, for a finite set X;

Ω is a 24-set, identified with the 24-set of Section 5;

$\mathcal{A} := \{\alpha_i \mid i \in \Omega\}$ is a set of vectors in $\mathbb{R}^\Omega$ which satisfy $(\alpha_i, \alpha_j) = 2\delta_{ij}$; we call $\mathcal{A}$ the *standard basis* of $\mathbb{R}^\Omega$;

$\pm 2\mathcal{A}$ is the *standard frame*;

for $S \subseteq \Omega$, $\alpha_S := \sum_{i\in S} \alpha_i$;

$M := \oplus_{i\in\Omega}\mathbb{Z}\alpha_i$, M_0 the even sublattice of M; note that M, M_0 are isometric to $\sqrt{2}\cdot SQ_{24}$, $\sqrt{2}\cdot L_{D_{24}}$, respectively;

$L_0 := M_0 + \sum_{S\in\mathcal{G}} \mathbb{Z}\left[\frac{1}{2} - \alpha_S\right]$; $\nu_i := -\frac{1}{4}\alpha_\Omega - \alpha_i$, $i \in \Omega$; note that $\nu_i - \nu_j \in M_0$;

$\Lambda := L_0 + \mathbb{Z}\nu_i$; note that this is independent of i; Λ is called the *standard Leech lattice* (see (9.2));

$L := M + L_0$, a Niemeier lattice (8.23) with root system A_1^{24};

ε_S is the linear map defined by $\alpha_i \mapsto (-1)^{i\cap S}\alpha_i$;

Σ_Ω is the permutation group on Ω, regarded as a monomial group with respect to $\mathcal{A}$, and $Mon(\Omega, \mathbb{Z})$ denotes the monomial group on the basis $\mathcal{A}$. with coefficients in the ring $\mathbb{Z}$; this group consists of matrices with entries 0 and ± 1 and is isomorphic to $\mathbb{Z}_2 wr \Sigma_\Omega$; the base group of this wreath product is $\{\varepsilon_S \mid S \in P\Omega\}$;

The *semiintegers* are the elements of $\frac{1}{2} + \mathbb{Z}$.

A lattice vector v has *type* n if $(v, v) = 2n$; the set of type n vectors in Λ is denoted Λ_n.

A *frame* is a set of 48 vectors in Λ_4, two of which are equal, opposite or orthogonal.

(9.2) Theorem. (i) *Λ is a Niemeier lattice with no roots.* (ii) *The minimal vectors are of type 2 and Λ_2 is the disjoint union of the sets*

$$\begin{aligned}
\Lambda_2^2 &= \left\{\tfrac{1}{2}\alpha_{\mathcal{O}}\varepsilon_S \mid \mathcal{O} \textit{ an octad, } S \in \mathcal{G}\right\} \\
&= \left\{\tfrac{1}{2}\alpha_{\mathcal{O}}\varepsilon_A | A \in PE(\mathcal{O}), \mathcal{O} \textit{ is an octad}\right\}, \\
\Lambda_2^3 &= \{\nu_i\varepsilon_S \mid i \in \Omega, S \in \mathcal{G}\}, \\
\Lambda_2^4 &= \{\pm\alpha_i \pm \alpha_j \mid i \neq j \textit{ in } \Omega\};
\end{aligned}$$

their cardinalities are $759 \cdot 2^7 = 97152$, $24 \cdot 2^{12} = 98304$ *and* $\binom{24}{2} 2^2 = 1104$ *and so* $|\Lambda_2| = 196560$.

Proof. (i) We have $L_0/M_0 \cong \mathcal{G}$ via $\alpha_S + M_0 \mapsto S$. Therefore, by (8.11), $det\, L_0 = 4$. Since $|\Lambda : L_0| = 2$, $det\, \Lambda = 1$. Using (8.15), it is easy to check that Λ is integral and even. Notice that $\frac{1}{2}\alpha_S \in \Lambda$ iff $S \in \mathcal{G}$ (hint: study $(\frac{1}{2}\alpha_S, \frac{1}{2}\alpha_T)$).

Now to prove that Λ has no roots. Clearly, M_0 contains no roots, so suppose that r is a root and $r \in \frac{1}{2}\alpha_S + M_0$, for some $S \in \mathcal{G}$. Since $|S| \geq 8$, $(r, r) \geq 2 \cdot \frac{1}{4} 8 = 4$, a contradiction. If $r \in \nu_i + L_0$, every coordinate is nonzero and is at least 1/4, whence $(r, r) \geq 1/16 \cdot 2 \cdot 24 = 3$, a final contradiction.

(ii) It is easy to see that $M_0 \cap \Lambda_2$ is the set Λ_2^4. Let $x \in \Lambda_2 \cap L_0 \backslash M_0$. For some $S \in \mathcal{G}$, $S \neq 0$, $x = \sum_{i \in S} c_i \alpha_i + v$, where v is a vector supported in $\Omega \backslash S$ and where the c_i are semiintegers. We have $4 = (x, x) \geq \frac{1}{4} \cdot 2 \cdot |S| + (v, v)$, whence S is an octad, $v = 0$ and each $c_i = \pm\frac{1}{2}$. Thus, there is $A \in P(\mathcal{O})$ such that $x = \frac{1}{2}\alpha_{\mathcal{O}}\varepsilon_A = \frac{1}{2}\alpha_{\mathcal{O}} - \alpha_A$. Let $i \notin \mathcal{O}$. Then integrality of (x, ν_i) implies that $\frac{1}{4} \cdot 2 \cdot |A| \in \mathbb{Z}$, i.e., that A is even. So, such an x is in Λ_2^2. Finally, let $x \in \Lambda_2$ have shape $\frac{1}{4}\sum c_i \alpha_i$, where the c_i are odd integers. Then, $(x, x) = 4$ implies that $\sum c_i^2 = 32$, whence there is an i such that $c_i = \pm 3$ and $c_j = \pm 1$ for $j \neq i$, i.e., $x = \nu_i \varepsilon_A$, for some $A \in P\Omega$. Replacing x by $-x$ if necessary, we may assume that a coefficient c_i is $\frac{3}{4}$. Then, $x = \nu_i \varepsilon_A = \nu_i - \frac{1}{2}\alpha_A$ is in Λ iff $A \in \mathcal{G}$. □

(9.3) Remarks. (*i*) $Aut\,(L)$ preserves M, the root sublattice, whence $Aut\,(L)$ is in $2wr\Sigma_{24}$ and it contains all diagonal maps. Since $Aut\,(L)$ maps to the subgroup of Σ_{24} preserving the Golay code via its action on $\frac{1}{2}M/M$, $Aut\,(L) = 2wrM_{24}$.

(ii) Clearly, $Aut\,(L_0)$ contains the natural M_{24} subgroup of $Aut\,(L)$ and it contains the diagonal map ε_A iff for all $S \in \mathcal{G}$, $\frac{1}{2}\alpha_S\varepsilon_A = \frac{1}{2}\alpha_S - \alpha_{S \cap A}$ is in L_0 iff all $S \cap A$ are even sets iff $A \in \mathcal{G}$. So, the subgroup of $Aut\,(L)$ which preserves L_0 is the natural $\mathcal{G} : M_{24}$.

(iii) We prove that the restriction of $\mathcal{G}{:}M_{24}$ to L_0 is onto $Aut\,(L_0)$. Consider the fours group L_0^*/L_0. The three subgroups of order 2 correspond to Λ (no roots), L (an even lattice with roots) and $L_0 + \mathbb{Z}\left(\frac{1}{4}\alpha_\Omega\right)$, which is odd integral. Thus, the action of $Aut\,(L_0)$ on this fours group must be trivial, whence $Aut\,(L_0)$ preserves L, and so the restriction is an isomorphism onto.

(iv) From (iii), we conclude that $\mathcal{G}{:}M_{24}$ is in $Aut\,(\Lambda)$.

(9.4) Theorem. *The theta function of a Niemeier lattice without roots begins* $\vartheta(z) = 1 + 196560q^2 + 16773120q^3 + 398034000q^4 + \cdots$.

Proof. This is a well-known consequence of standard results from modular form theory. For instance, see [Serre]. We already know from (9.3) that the coefficient at q^2 is 196560. □

(9.5) Theorem. *Let* Λ *be a Niemeier lattice without roots. Then:*

(i) *The intersection of the coset with* $\Lambda_0 \cup \Lambda_2 \cup \Lambda_3 \cup \Lambda_4$ *consists of exactly one of the following:* (a) $\{0\}$; (b) *a vector of type* 2 *and its negative;* (c) *a vector of type*

3 and its negative; (d) *a set of vectors of type* 4*, two of which are equal, negatives or orthogonal. We say that a coset has type* $k \in \{0,2,3,4\}$ *iff it contains a vector of type* k.

(ii) *The cosets of type* 4 *have exactly* 48 *vectors of type* 4 *and these constitute a frame.*

Proof. We first show that The cosets given by the criteria (a), (b), (c) and (d) are distinct. In the case of a coset of type (d), it is trivial that there are at most 48 vectors of type 4 in such a coset, for a set of pairwise orthogonal vectors are independent. Using (9.4), we find that such cosets number at least hence exactly $1 + 196560/2 + 16773120/2 + 398034000/48 = 16777216 = 2^{24}$, and so conclude that we have accounted for all of them. An alternative to quoting (9.4), which requires modular form theory, is to show that there are at least 398034000 vectors of type 4 by exhibiting them, in the style of (9.3).

Let $X := \Lambda_0 \cup \Lambda_2 \cup \Lambda_3 \cup \Lambda_4$ and let $x, y \in X$, $x \neq y$. Then $(x-y, x-y) = x^2 - 2x.y + y^2 \leq 8 - 2x.y + 8$. Assuming $x - y \in 2\Lambda$, we get $(x-y, x-y) \geq 4 \cdot 4 = 16$, whence $x.y \leq 0$. It is also true that $x + y \in 2\Lambda$, and a similar argument gives $x.y \geq 0$ and so $x.y = 0$. We conclude that $(x \pm y)^2 = x^2 + y^2 = 16$ and so x and y have type 4, as x^2, $y^2 \leq 8$. Therefore, the coset $x + 2\Lambda$ meets X in a set of vectors of type 4, two of which are proportional or orthogonal. □

(9.6) Notation. T is a 4-set and Ξ is its associated sextet (5.10). For $T \in \Xi$, define $E_T \in End(V)$, $V = \mathbb{Q} \otimes_{\mathbb{Z}} \Lambda$, by $\alpha_j \mapsto 0$ if $j \notin T$ and $\alpha_j \mapsto \frac{1}{2}\alpha_T - \alpha_j$. Define $\xi := \sum_{T \in \Xi} E_T$ and $\eta_T := \xi \varepsilon_T$. Easily, one sees that ξ and η_T are involutions.

As matrices, each is a block sum of submatrices of shape $\frac{1}{2}\begin{pmatrix} -1 & 1 & 1 & 1 \\ 1 & -1 & 1 & 1 \\ 1 & 1 & -1 & 1 \\ 1 & 1 & 1 & -1 \end{pmatrix}$.

We write N_{24} for $\mathcal{G}{:}M_{24}$.

(9.7) For each T, $\eta_T \in Aut(\Lambda) \setminus N_{24}$.

Proof. Write $\Xi = \{T_1, \ldots, T_6\}$, $\eta_i = \eta_{T_i}$. It suffices to prove the statement for one i since the ratio $\eta_i^{-1}\eta_j = \varepsilon_{T_i + T_j}$ is in N_{24}. Also, it suffices to prove that $(\Lambda)\eta_i \leq \Lambda$ since η_i is orthogonal.

It is clear that a sum or difference of two vectors of the form $\alpha_k E_l$ lies in Λ, whence $(M_0)\eta_i \leq \Lambda$. Suppose $\mathcal{O}$ is an octad; its intersections with the T_k have constant parity, say p.

Suppose p is even. If $\mathcal{O} = T_{kl}$, $i \notin \{k, l\}$, then $\alpha_{\mathcal{O}}\eta_i = -\alpha_{\mathcal{O}} \in \Lambda$. We may assume that there is a 4-set J of indices such that $\mathcal{O}$ meets T_j in a 2-set iff $j \in J$. Then $\alpha_{\mathcal{O}}\eta_i = \alpha_{\mathcal{O}'} \in \Lambda$, where $\mathcal{O}'$ is the octad $T_J + \mathcal{O}$.

Finally, suppose p is odd. Then $\mathcal{O}$ meets each T_k in a 1– or a 3-set; without loss, $\mathcal{O} \cap T_i$ is a 3-set, $\mathcal{O} \cap T_j$ is a 1-set for $j \neq i$. Then, $\alpha_{\mathcal{O}}\eta_i = \nu_r \varepsilon_{\mathcal{O}}$, where $\{r\} = T_i \setminus \mathcal{O}$. So, in all cases, $\alpha_{\mathcal{O}}\eta_i \in \Lambda$. Since the octads generate the Golay code, this result, with $M_0\eta_i \leq \Lambda$, proves that $L_0\eta_i \leq \Lambda$. Finally, we use the fact, proved above, that $\nu'\eta_i \in \Lambda$, where $\nu' = \nu_r\varepsilon_O \in \nu_r + L_0$, to conclude that η_i maps Λ onto Λ. □

(9.8) Lemma. *The subsets Λ_2^2, Λ_2^3 and Λ_2^4 each determine the standard frame F in the sense that*

(a) *$\Lambda_2^4 + 2\Lambda$ spans the subspace $M_0 + 2\Lambda/2\Lambda$ and if we define $A := \{x \in M_0 + 2\Lambda \mid x$ is a vector of type 4 which is a sum of two elements of $\Lambda_2^4\}$ and $B := \{x \in A | (x, A) \subseteq 4\mathbb{Z}\}$, then $B = F$;*

(b) *$F + 2\Lambda$ spans the radical modulo 2 of $L_0 = span\, \Lambda_2^2$;*

(c) *$\Lambda_2 \backslash \Lambda_2^3 = \Lambda_2^2 \cup \Lambda_2^4$, which spans L_0 and so determines F as in* (b).

Consequently, the stabilizer in $Aut(\Lambda)$ of any of the sets Λ_2^2, Λ_2^3 or Λ_2^4 is the stabilizer of F in $Aut(\Lambda)$.

Proof. Verifications of the statements in (a), (b) and (c) are straightforward. Clearly, they imply that the three stabilizers lie in the stabilizer of F, which clearly stabilizes each Λ_2^k. □

(9.9). Proposition. *A subgroup of $Aut(\Lambda)$ containing N_{24} properly (e.g., $Aut(\Lambda)$) operates transitively on the set of* 196560 *type* 2 *vectors. (The action is not primitive since the sets $\{x, -x\}$ form blocks; see Section 10 for a discussion of primitivity).*

Proof. Let $X = \Lambda_2$. There are three orbits of $N := N_{24}$ on X, namely Λ_2^2, Λ_2^3 and Λ_2^4. Let H be any subgroup containing N properly; by (9.7), H exists. By (9.8), H stabilizes none of these three sets, so H is transitive. □

(9.10) Proposition. *The stabilizer in $Aut(\Lambda)$ of the standard frame $\{2\alpha_i \mid i \in \Omega\}$ is just N_{24}.*

Proof. Let S be the stabilizer of the frame F in $Aut(\Lambda)$. Then, S stabilizes both $\frac{1}{2}F$ and $\{x \in \Lambda \mid (x, F) \subseteq 2\mathbb{Z}\}$, which is L_0. From (9.3.iii), $Aut(L_0) = N_{24}$. Since S contains N_{24}, they are equal. □

(9.11) Remark. In analogy with Sections 5 and 6, what would follow here is the definition of labeling of a frame, the classification of labelings on a given frame and the proof that $Aut(\Lambda)$ acts transitively on frames. These arguments are placed in an appendix to this chapter. Instead we present here a more direct proof of transtivity on frames.

(9.12) Definition. An *overweight A_1^{24}-lattice* is a lattice P whose dual P^* is a sublattice of index 4 such that

(9.12.1) $P/P^* \cong 2 \times 2$;

(9.12.2) the three sublattices between P^* and P are isometric to the Leech lattice, the Niemeier lattice $\mathcal{N}_{A_1^{24}}$ and the odd integral lattice $L_0 + \mathbb{Z} - \frac{1}{4}\alpha_\Omega$, in the notation of (9.1).

Note that $L+\Lambda$ contains exactly two even, integral unimodular lattices of index 2, namely L and Λ. Since P has the corresponding property, it is easy to prove that P is isometric to $L+\Lambda$ by noticing that P^* is isometric to $L \cap \Lambda = L_0$.

(9.13) Theorem. *Containment induces bijections between the following three sets of lattices in $\mathbb{Q}^\Omega := \mathbb{Q} \otimes M$: Leech lattices with a choice of frame, Niemeier lattices with root systems of type A_1^{24}, and overweight A_1^{24}-lattices. Consequently, the orthogonal group on $\mathbb{Q}^\Omega$ is transitive on sublattices of $\mathbb{Q}^\Omega$ of these three types in* (9.12). *In particular, it is transitive on Leech lattices.*

Proof. The containment statement is clear from the constructions of these lattices and uniqueness of the Golay code.

Let P and Q be such and let E and F be the unique frames in their Leech sublattices such that $\frac{1}{2}E \subseteq P$ and $\frac{1}{2}F \subseteq Q$. There is an orthogonal transformation g which carries E to F and g carries P to an overweight A_1^{24} lattice containing F.

The set of overweight A_1^{24} lattices containing F is in bijection with the set of binary Golay codes in $\mathbb{F}_2^{24}$. The exact correspondence is as follows. If $\langle F \rangle$ denotes the $\mathbb{Z}$-span of F and Y is an overweight A_1^{24}-lattice containing F, then Y is between $\langle F \rangle$ and $\langle F \rangle^*$. The double basis F of $\mathbb{Q}^\Omega$ gives a preferred basis in $\langle F \rangle^*/\langle F \rangle$, the image of F, and $Y/\langle F \rangle$ is a binary Golay code.

By the uniqueness result for binary Golay codes (5.29), there is a permutation of the basis which carries $P^g/\langle F \rangle$ to $Q/\langle F \rangle$. Such a permutation comes from a monomial transformation on $\mathbb{Q}^\Omega$ with respect to the double basis F. The required transitivity follows. □

(9.14) Corollary. (i) *Aut* (Λ) *is transitive on the set of frames and on the set of vectors of type 4; its order is* $2^{22}3^9 5^4 7^2 11{\cdot}13{\cdot}23$; (ii) *Aut* (Λ) *is transitive on the vectors of type* 2 *and the stabilizer of a vector of type* 2 *has order* $2^{18}3^6 5^3 7{\cdot}11{\cdot}23$.

Proof. (i) (9.10) and (9.13). (ii) Use (i) and (9.9) (note that (9.13) could be used instead of (9.3) to get (9.9)). □

(9.15) Lemma. *Let $P \in Syl_{23}(Aut(\Lambda))$, $N := N(P)$. Then, $N \cong 23{:}11 \times 2$ and N stablilizes a frame.*

Proof. By Sylow's theorem for $G = Aut(\Lambda)$, we may take $P = \langle g \rangle$ as a subgroup of the natural M_{24} in G. Let L be the fixed point sublattice. Rational canonical form for g shows that L has rank 2 and contains $x = \frac{1}{4}\left(-3\alpha_i + \sum_{j \neq i} \alpha_j\right)$ where i is the unique index fixed by g and $y = \frac{1}{4}\left(5\alpha_i + \sum_{j \neq i} \alpha_j\right)$. Note that $x{.}x = 4$, $y{.}y = 6$ and $x{.}y = 1$, so that L', ths span of x and y, has determinant 23. Consequently, $L = L'$, for any lattice between L' and L has determinant of the form $23/m^2$, where m is its index in L; see (8.16). Now, N stabilizes L, whose set of elements of type 4 is $\pm(x-y) = 2\alpha_i$ (proof: say $ax+by$ has type 4;

then $8 = a^2 4 + 2ab + b^2 6$, whence $(a, b) = \pm(1, -1)$). From (9.10), we conclude that N lies in the stabilizer of the frame containing $x - y$. By studying N_{24} (see (5.41)), we get that $N = N_{M_{24}}(P) \times \langle \varepsilon_\Omega \rangle$; the first factor is described in (1.18) and $\varepsilon_\Omega = -1$. □

(9.16) Exercises. (9.16.1) If M_{22} is embedded in Σ_n, $n \geq 22$ (Hint: simplicity of M_{22} and (1.4) will eliminate all $n < 22$ dividing $|M_{22}|$ but 12; then use the $2^4{:}Alt_6$ subgroup of $G(\mathcal{O})$ defined as the stabilizer of two points in $\Omega + \mathcal{O}$).

(9.16.2) Let $\mathcal{U}$ be the subspace of Golay sets which avoid the 2 points of Ω fixed by M_{22}; if $\mathcal{U} : M_{22} \cong 2^{10} : M_{22}$ is embedded in Σ_n, then $n \geq 44$ (Hint: show that $\mathcal{U}$ is an irreducible module for M_{22} by restricting to a group of order 11, then use (9.16.i)).

(9.17) Proposition. *N_{24} is a maximal subgroup of $Aut(\Lambda)$ and is not normal.*

Proof. The last statement is trivial since otherwise N_{24}, having odd index, would contain all Sylow 2-subgroups, whereas the involution η_T of (9.7) is not in N_{24}.

Suppose $N := N_{24} < H \leq G := Aut(\Lambda)$. Then, (9.9) implies that H is transitive on X, the set of type 2 vectors. On X, N has three orbits, with stabilizers $N_4 \cong 2^{10} M_{22}{:}2$, $N_3 \cong M_{23}$ and $N_2 \cong 2^5 2^4\, Alt_8$. So, if $x \in X$, $|H_x|$ is divisible by $m := 2^{18} 3^2 5 \cdot 7 \cdot 11 \cdot 23$. In the group G_x, of order $2^{18} 3^6 5^2 7 \cdot 11 \cdot 23$ (see (9.14)), H_x is a subgroup of index i dividing $3^4 5^2$. Since N_2 contains $23{:}11$ and -1 does not stabilize x, (9.15) implies that the full Sylow 23 normalizer in G_x must have shape $23{:}11$ and so i is $1 (mod\, 23)$, whence $i = 1$ or $3^4 5^2$.

Suppose $i = 3^4 5^2$. Then, H_x contains N_4 with index 23. Since $N_4/O_2(N_4) \cong M_{22}$ is simple, (9.16.ii) implies that $O_2(N_4)$ is normal in H_x. Since H_x permutes the eigenspaces of $O_2(N_4)$, H_x stabilizes the standard frame and so is in N_{24}. Since $M_{22}{:}2$ is maximal in M_{24}, which has no embedding in Σ_{12}, the facts $N_4 \leq H_x \leq N_{24}$ and $|H_x{:}N_4| = 23$ lead to a contradiction. Therefore $i = 1$ and so H contains G_x. Since H is transitive on Λ_2, $G = G_x H \leq HH = H$, contradiction. □

We prove transitivity on vectors of type 3.

(9.18) Lemma. *Let x have type 3. There is $y \in \Lambda_4$ such that $x.y$ is odd.*

Proof. Consider the annihilator W of $x + 2\Lambda$ in $\Lambda/2\Lambda$. It has rank 23 and has the property that W contains exactly $2^{22} - 1$ nonzero singular vectors (because, for $y \in W$, $\frac{1}{2}(x+y)^2 \equiv \frac{1}{2}x^2 + \frac{1}{2}y^2 \equiv 1 + \frac{1}{2}y^2 (mod\, 2)$). The cardinality of the set of cosets represented by elements of order 4 is $|\Lambda_4|/48 = 8292375 > 2^{22} - 1 = 4194303$. □

(9.19) Lemma. *Let y have type 4 and let $A(y) := \{x \in \Lambda \mid x$ has type 3 and $x.y$ is odd $\}$. Then, $Stab_{Aut(\Lambda)}(y)$ has eight orbits on $A(y)$. If $y = 2\alpha_r$, these orbits are represented by the eight vectors of shape $\pm\frac{1}{4}(5\alpha_i + \sum_{j \neq i} \alpha_j)$ for $r = i$ and* *some* $r \neq i$; $\pm\frac{1}{4}(-3\alpha_j - 3\alpha_k - 3\alpha_l + \sum_{i \neq j,k,l} \alpha_i)$, *for* $r \in \{j,k,l\}$, *and some* $r \notin \{j,k,l\}$.

Proof. We may assume (9.14.i) that $y = 2\alpha_i$. Since $x{\bullet}y$ is odd, x is in the coset $\nu_i + L_0$. Since $x{\bullet}x = 6$, the absolute values of the 24 coordinates must be $\frac{1}{4}(5^1 1^{23})$ or $\frac{1}{4}(3^3 1^{21})$. Since $x \pm \nu_i \in \Lambda$ for all i, it follows from high transitivity of M_{24} that the eight vectors above represent the orbits of $Stab_{Aut(\Lambda)}(y) \cong 2^{11}{:}M_{23}$. □

(9.20) Theorem. *$Aut(\Lambda)$ acts transitively on the set of* $16773120 = 2^{12}3^2 5.7.13$ *vectors of type* 3.

Proof. We show that the eight elements of (9.19) lie in a single $Aut(\Lambda)$-orbit. Let i, j, k, l be distinct indices, let Ξ be the sextet associated to this four set and for $T \in \Xi$, and define the automorphisms η_T as in (9.6). A straightforward calculation shows that, for $T = \{i,j,k,l\}$, η_T carries the vector $\frac{1}{4}\alpha_\Omega + \alpha_i$, with coefficients $\frac{1}{4}(51^{23})$ to one with coefficients $\frac{1}{4}(3^3 1^{21})$ iff $i \in T$. The action of N_{24}, transitivity of $Aut(L)$ on vectors of type 4 and (9.18) finish the proof. □

(9.21) Definition (The Conway Groups). For $n = 0$, 2 or 3, define Co_n as the stabilizer in $Aut(\Lambda)$ of a vector of type n. Since there is one orbit on vectors of type n, Co_n is defined up to conjugacy in $Aut(\Lambda) = Co_0$. We define $Co_1 := Co_0 / \{\pm 1\}$. Sometimes, Co_n (the *n-th Conway group*) is denoted $\cdot n$ (pronounced, "dot n").

Also, we denote by $\cdot abc$ the stabilizer of a *triangle of type* a, b, c, i.e., a set of lattice vectors which sum to 0 and have respective types a, b and c.

For certain triples (a, b, c), there is just one orbit on such triangles, so the group $\cdot abc$ is determined up to conjugacy. For example, (9.19) implies that this is so for triangles of type $43k$, for $k = 2$, 4, 10 or 12; but transitivity does not hold for $k = 6$ or 8. See Section 10.

(9.22) Proposition. *Co_0 is perfect and Co_1 is simple.*

Proof. Let K be a normal subgroup of $G = Co_0$ which is minimal subject to containing -1 but not being in the center. The orbits of K on Λ_2 have common length, say $d > 1$, where $d | 2^4 3^2 5.7.13$. Let P be a p-subgroup of G. If 23 divides $|\,Aut(P)\,|$, (1.17) implies that $|P| = 2^m$, for some $m \geq 11$.

If 23 does not divide the order of K, (1.10), (9.15) and the last remark imply that K is a 2-group, of order 2^m, for some $m \geq 11$. Since KN_{24} is a subgroup of G and since N_{24} has odd index, $K \leq N_{24}$, whence $K \leq O_2(N_{24})$. By (5.41), $K = O_2(N_{24})$. The 1-dimensional spaces $\mathbb{Q} \otimes \Lambda$ afford 24 distinct linear characters of K and their intersections with Λ are generated by the vectors $\{\pm 2\alpha_i \mid i \in \Omega\}$. It follows that the standard frame is stable under G since K is normal in G, in contradiction to (9.10) and (9.17).

We now have that K contains a Sylow 23-group of G. Minimality of K, (9.15) and (1.12) imply that K contains a Frobenius group of shape 23:11. The intersection $K \cap N_{24}$ therefore contains a Sylow 23-normalizer in the standard M_{24}-subgroup and is normal in N_{24}. Simplicity of M_{24} and the structure of $\mathcal{G}$ as a module for M_{24} (see (5.41)) imply that K contains N_{24}. Since N_{24} is maximal in G and is not normal (9.17), $K = G$. If G is not perfect, simplicity of $G/\{\pm 1\}$ implies that $G = G' \times \{\pm 1\}$, contrary to (5.41). □

(9.23) Exercise. Determine the centralizer in N_{24} of η_T, the isomorphism type of the group $\langle \eta_T T \in \Xi \rangle$ (9.6) and its normalizer in N_{24}.

(9.24) Exercise *(Ternary Golay code construction of the Leech lattice).* This is an analogue of the tetracode construction given for the E_8-lattice (8.22.2).

(9.24.1) Give a construction of an even unimodular lattice of rank 24 by starting with a root lattice for A_2^{12} and the ternary Golay code $\mathcal{TG}$, following the ideas of (8.22.2). This lattice has roots and is the Niemeier lattice with root system A_2^{12}.

(9.24.2) Enlarge this lattice by index 3 to get an "overweight A_2^{12} lattice", one which contains a rootless Niemeier lattice of index 3. Show that we get a subgroup of $Aut(\Lambda)$ isomorphic to $\mathcal{TG}: Aut(\mathcal{TG}) \cong 3^6{:}2{\cdot}M_{12}$. Formulate and prove a transitivity result analogous to transitivity of $Aut(\Lambda)$ on frames.

(9.24.3) Generalize to overweight versions of other Niemeier lattices.

Appendix 9A. Labeling Analysis for Frames in a Leech Lattice

See Remark (9.11) for orientation to this section. Though we may use the uniqueness of the binary Golay code, a labeling analysis here is more subtle since here we have to deal with integers modulo 4, not 2.

(9A.1) Notation. A *frame* is a set of 48 vectors of type 4 in Λ such that two of which are equal, negatives or orthogonal; a frame is a double basis. A *labeled frame* is a frame F with a set map $\Omega \to F$, called a *labeling*, whose image is a basis. We let $\mathcal{G}$ be the standard Golay code and let $\Lambda(F, \mathcal{G})$ be the lattice defined as in (9.1), using $2\alpha_i$ for the image of i under our labeling.

(9A.2) Remark. For the next result, we follow the ideas in the proof that M_{24} is transitive on sextets, given in (5.30). Given two such, $\Lambda := \Lambda(F, \mathcal{G})$ and $\Lambda' := \Lambda(F', \mathcal{G}')$, there is a unique invertible linear transformation g such that $F^g = F'$ and $\mathcal{G}^g = \mathcal{G}'$. Then, $\Lambda(F, \mathcal{G})^g = \Lambda(F', \mathcal{G}')$. Also, g is orthogonal.

(9A.3) Proposition. *Let L be a Leech lattice and F a frame in L. Then, there is a labeling of F such that $L = \Lambda(F, \mathcal{G})$.*

Proof. To begin, we take any labeling of F. Observe that L is between M and M^*, where M is the $\mathbb{Z}$-span of F. Since M is isometric to $\sqrt{8} \cdot SQ_{24}$, $det\, M = 2^{72}$

and $M^*/M \cong \mathbb{Z}_8^{24}$. Let $A = M^*/M$, $A_k = \{x \in A | 2^k x = 0\}$ and define M_k by $M_k/M = A_k$. We have $M^* = \oplus\mathbb{Z}\left(\frac{1}{2}\alpha_i\right)$, where the choices $2\alpha_i$ represent a labeling of the frame. For $S \in P\Omega$, define $M_{k,S} := (\oplus_{i\in S}\mathbb{Q}\alpha_i) \cap M_k$ and write $M_{k,i}$ for $M_{k,\{i\}}$. Since L has no roots, $L \cap M_{1,i} = M_{0,i}$, for all i. Let $L(k) := L \cap M_k$. Then $L(1)/M_0$ has rank at most 23. Since F is a frame in L, the vectors $\frac{1}{2}(x \pm y)$ are in L, for $x, y \in F$. Therefore, $L(1) = \{\sum c_i\alpha_i \mid \text{the } c_i \text{ are integers}$ and $\sum c_i$ is even$\}$. Notice that there is a pairing of $L/L(1) \times L(1) \to \mathbb{Z}/2\mathbb{Z}$ coming from reducing the usual inner product modulo 2. Let $v = \sum c_i\alpha_i \in L \backslash L(2)$, and let I be the subset of Ω where the c_i are not in $\frac{1}{2}\mathbb{Z}$. If $i \in I$, $j \notin I$, then $(v, \alpha_i + \alpha_j) \notin \mathbb{Z}$. We conclude that $I = \Omega$.

We now have $L/L(2)$ of order 2, and $L(1)/M$ of order 2^{23}. Consider $L(2)/L(1)$, which is naturally isomorphic to a code, C, in $M_2/M_1 \cong 2^{24}$. Since $det\, L = 1$, $|L : M| = 2^{36}$, whence C has dimension 12.

We may add vectors of the form $\alpha_i \pm \alpha_j$ to v (as above) to arrange for all c_i to be $\pm\frac{1}{4}$ or all but one to be $\pm\frac{1}{4}$ and the last to be $\pm\frac{3}{4}$. Since $(v, v) \geq 4$, only the second possibility may occur. For such a v, let $E(v) = \{i \mid c_i = -\frac{1}{4} \text{ or } \frac{3}{4}\}$.

We claim that C is a self-orthogonal code of minimum weight 8. Let d be a nonzero weight and $c \in C$ of weight d. Then $L(2)$ contains a vector u of shape $\sum c_i\alpha_i$, where the c_i are semiintegers at c and are otherwise integers. We may add to u elements of the form $+\alpha_i \pm \alpha_j$ to arrange for all c_i to be $\pm\frac{1}{2}$ or 0. If so, $(u, u) = d/2$, whence $d \geq 8$. Since $dim\, C = 12$, C is a Golay code. By (5.30), C is equivalent to the standard Golay code, $\mathcal{G}$ (a permutation matrix on $\mathbb{R}^\Omega$ effecting this tranformation of codes stabilizes M); that is, we may reidentify Ω (9.1) with the 24-set of Section 5.

We need a further refinement of this labeling to get all $\frac{1}{2}\alpha_S$ in L, where $S \in \mathcal{G}$; at this point, this statement is true only up to possible sign changes (the procedure of the last paragraph, adding various $\pm\alpha_i \pm \alpha_j$ will not get $\frac{1}{2}\alpha_S$ unless the number of c_i equalling $-\frac{1}{2}$ is even). Let $\mathcal{B}$ be a basis of $\mathcal{G}$ and for each $B \in \mathcal{B}$, let $N := N(B)$ be the subset of B such that $\frac{1}{2}\alpha_B\varepsilon_N$ is in L. There is $T \in P(\Omega)$ such that $|T \cap B| \equiv |N(B)| (mod\, 2)$, for all $B \in \mathcal{B}$. If we replace the labeling map by its composite with ε_T, we then have that each $N(B)$ is even. Then, for each $B \in \mathcal{B}$, we get $\frac{1}{2}\alpha_B \in L$ by the earlier procedure. The equation $\frac{1}{2}\alpha_S + \frac{1}{2}\alpha_T = \frac{1}{2}\alpha_{S+T} + \alpha_{S\cap T}$ then implies that all $\frac{1}{2}\alpha_S$ are in L.

Now, let $S \in \mathcal{G}$ and consider $\left(v, \frac{1}{2}\alpha_S\right)$, where v is the vector described at the end of the last paragraph. It is $\equiv 24\cdot\frac{1}{4}\cdot 2 + \frac{1}{4}2\cdot|E(v) \cap S|\, (mod\, \mathbb{Z})$. Reverse the labeling at all indices in $E(v)$; then all $\left(v, \frac{1}{2}\alpha_S\right) \in \mathbb{Z}$. Since $\left(L(2), \frac{1}{2}\alpha_S\right) \leq \mathbb{Z}$, it follows that $\frac{1}{2}\alpha_S \in L^* = L$. We conclude that L is contained in $\Lambda(F, \mathcal{G})$. Since both lattices are unimodular, $L = \Lambda(F, \mathcal{G})$. □

Chapter 10. Subgroups of the Conway Groups; the Simple Groups of Higman-Sims, McLaughlin, Hall-Janko and Suzuki; Local Subgroups; Conjugacy Classes

We describe the involvement in Co_0 of the simple groups of Hall-Janko, Suzuki, McLaughlin and Higman-Sims. In addition, we determine conjugacy classes of elements of small order and the structure of several local subgroups (recall that a *p-local subgroup*, for a prime p, is the normalizer of a nonidentity p-subgroup, and a *local subgroup* is a p-local subgroup for some prime p). Appendix $10C$ contains Conway's original table of triangle and other stabilizers, along with additional material. Our treatment of the conjugacy classes is original and our treatment of the above simple groups does involve some revision.

Due to the increased complexity of the groups, we need to quote somewhat nontrivial facts about the finite simple groups. To keep our text relatively self-contained, we hold these to a minimum and use elementary arguments whenever possible. We shall need some properties and terminology of the classical groups, Chevalley groups and finite geometries, beyond the elementary material in Section 2, for instance, a few things from the theory of parabolic subgroups and stabilizers of subspaces in the standard modules. It seems best to refer the reader to a number of standard texts, e.g. [Car], [Artin], [Dieu], [Wie]. We also need results from the classification of finite simple groups; see Appendix $10B$. The latter will be needed to identify sections of Co_0 as simple groups, including those of HS, McL, HJ and Suz. For general background, see [Go68].

Throughout this section, G denotes $Co_0 = Aut(\Lambda)$. The notation $X \sim_H Y$, or just $X \sim Y$, means X and Y are conjugate subsets or elements of a group H. We say that an element z of a group X is *weakly closed* in a subgroup Y if $Y \cap \{z^g \mid g \in X\} = \{z\}$. Recall that a *triangle of type abc* is a triple of vectors in the Leech lattice, of types a, b and c, which sum to 0.

(10.1) Notation. Recall that in (9.20), we proved transitivity on Λ_3. We use some notations like those of (9.18) and (9.19). Here are two elements of type 3.

(10.1.1) $$x' = \frac{1}{4}\left(5\alpha_i + \sum_{j \neq i} \alpha_j\right);$$

(10.1.2) $$x'' = \frac{1}{4}\left(3\alpha_j - 3\alpha_k - 3\alpha_l + \sum_{i \neq j,k,l} \alpha_i\right).$$

(10.2) Exercises. Let $x \in \Lambda_3$ and $y \in \Lambda_2$. Then, $x.y \in \{-4,-2,-1,0,1,2,4\}$. For k in this set, define $A_k(x) := \{y \in \Lambda_2 | x.y = k\}$. Recall the notations of Section 9, especially (9.1).

(i) Prove that $|A_{-3}(x)| = 552$ and enumerate the 253 triangles of type 223 containing x by surveying the vectors of type 2. Furthermore, show that

(10.2.1) if $x = x'$ (10.1.1), $A_{-3}(x)$ has the form:

23 vectors $-\nu_j$, $j \neq i$ and 253 vectors $\nu_i\varepsilon_S$, where S is a 16-set not containing i; in any triangle of type 223, the leg in $\Lambda(2)$ is the negative of the sum of one of these and x, so we get the 276 others, 23 of shape $-\alpha_{ij}$, $j \neq i$, and 253 of shape $\frac{1}{2}\alpha_{\mathcal{O}}$, where $\mathcal{O}$ is an octad containing i.

Determine $A_{-3}(x)$ in the following additional cases.

(10.2.2) $x'' = \frac{1}{4}\left(3\alpha_j - 3\alpha_k - 3\alpha_l + \sum_{i \neq j,k,l} \alpha_i\right)$, (10.1.2);

(10.2.3) $x = -\frac{1}{2}\alpha_{\mathcal{D}}$, for a dodecad $\mathcal{D}$;

(10.2.4) $x = -\frac{1}{2}\alpha_{\mathcal{O}} - \alpha_{ij}$, $i \in \mathcal{O}$, $j \notin \mathcal{O}$, for an octad $\mathcal{O}$.

(ii) Prove that $|A_{-2}(x)| = 11178 = 2.3^5 23$ and count triangles of type 233 containing x by surveying the vectors of type 2. Furthermore, show that

(10.2.5) $A_{-2}\left(x''\right)$ has the form:

1288 vectors $\nu_i\varepsilon_S$, where S is a dodecad avoiding i; 4048 = 23.176 vectors $\nu_j\varepsilon_S$, for $j \neq i$ and S a 16 set containing i but not j; 5313 = 253.21 vectors $-\frac{1}{2}\alpha_{\mathcal{O}}\varepsilon_T$, where $i \in \mathcal{O}$ is an octad and T meets $\mathcal{O}$ in a 6-set containing i; 506 vectors $-\frac{1}{2}\alpha_{\mathcal{O}}$, where $\mathcal{O}$ is an octad avoiding i; 23 vectors $-\alpha_i + \alpha_j$, $j \neq i$.

Also, determine $A_{-2}(x)$ in each case (10.2.3 and 4).

(iii) Use (i) and (ii) to study action of G on $\Lambda_3 \times \Lambda_2$. Deduce transitivity of G on triangles of type 223 and type 233. Use transitivity on vectors of type 3 (9.20). Suppose we are dealing with triangles of type 223. Let x be a vector of type 3 and $N := N_{24}$; we want transitivity of G_x on $A_{-3}(x)$. If $x = x'$, the group $N_x := G_x \cap N$ has orbits of lengths 23 and 253 on $A_{-3}(x)$. If x has another shape as in (10.2.3 or 4), we get fusion of these two orbits to a single orbit of length 276, for example, if $x = -\frac{1}{2}\alpha_{\mathcal{D}}$, $\mathcal{D}$ a dodecad, and $y = -\alpha_{\mathcal{O}}$, for an octad $\mathcal{O}$ such that $|\mathcal{O} \cap \mathcal{D}| = 6$; in this case, $N_x \cong 2 \times M_{12}$ and its orbit containing y has length $|M_{12}|/|\Sigma_6| = 132$ (6.15.iii). Similar analysis works for triangles of type 233.

(10.3) Theorem. *Co_3 is simple of order $2^{10}3^7 5^3 7 \cdot 11 \cdot 23$ and is a doubly transitive group on the set of 276 triangles of type 223 containing our fixed vector of type 3.*

Proof. From (9.20) and (10.2.3), there are 276 such triangles. Let Δ be this set of triangles. The set of legs other than x span $\mathbb{Q} \otimes \Lambda(10.2.i)$ and so Co_3 acts faithfully on Δ. Let K be a minimal normal subgroup. Every orbit of K on these $276 = 2^2 3.23$ triangles has the same length, say $d > 1$. If $(23, |K|) = 1$, we get a contradiction as in the proof of (9.22). So, $23 \mid |K|$ and $23 \mid d$. From (9.15) and the fact that Co_3 contains M_{23} (take $x = \frac{1}{4}\left(5\alpha_i + \sum \alpha_j\right) \in \Lambda_3$ to see this), we know that a Sylow 23-normalizer in K has shape 23:11 or 23:11 × 2; since the index is 1 (mod 23) the latter possibility is out and then the Fratttini argument gives $K = Co_3$. □

We next develop some information to prove simplicity of the triangle stabilizer .223, which turns out to be isomorphic to the simple McLaughlin group McL; see (10.5).

(10.4) Lemma. *Let g be an element of order* 11 *in* $M_{24} \leq N_{24} \leq G = Co_0$.

(i) The centralizers of g in these respective groups have shapes:

(10.4.1) $11, \quad 11 \times 2^2, \quad 11 \times Dih_{12}$.

(ii) The respective normalizers are:

(10.4.2) $11{:}10, \quad \left[11{:}5 \times 2^2\right]{:}2 \cong [11{:}10 \times Dih_8]\,\frac{1}{2}, \quad [11{:}5 \times Dih_{12}]{:}2$

(in the latter two cases, an outer involution induces outer automorphisms on each of the two factors).

(iii) An element of order 3 in $C_G(g)$ has 0 fixed point sublattice in Λ.

Proof. (i) The Golay sets fixed by an element of order 11 are $\emptyset$, Ω and a complementary pair of dodecads. Therefore, the normalizer lies in a natural $M_{12}{:}2$-subgroup. Now, see (6.19.ii) and use the Frattini argument. The structure of the N_{24}-normalizer follows from (i) and the Frattini argument. (iii) The fixed point sublattice L of the particular element α of order 11 (6.21) has rank 4. We claim that it is spanned by ν_i, $\frac{1}{2}\alpha_{\mathcal{D}}$, $\frac{1}{2}\alpha_{\mathcal{D}+\Omega}$ and α_{ij} (here, $i \in \mathcal{D}$ and $j \in \mathcal{D} + \Omega$ are the points of the two dodecads fixed by α). The sublattice M spanned by these four vectors has determinant 121 and so is a direct summand of Λ; this proves the claim. We now look for vectors of type 2 in $L = M$ and find that there are exactly these 12:

(10.4.3) (a) $\pm\alpha_i \pm\alpha_j$;

(b) $\nu_i g,\ g \in \langle \varepsilon_{\mathcal{D}}, \varepsilon_{\mathcal{D}+\Omega} \rangle$;

(c) $\nu_j g,\ g \in \langle \varepsilon_{\mathcal{D}}, \varepsilon_{\mathcal{D}+\Omega} \rangle$;

We make the critical observation that for the equivalence relation on pairs of vectors (x, y) as in (10.4.3) generated by the condition that $x - y \in \Lambda_4$, the equivalence classes are just

(10.4.4) (a′) $\pm\alpha_i \pm \alpha_j$;

(b′) $\pm\nu_i,\ \pm\nu_j\varepsilon_{\mathcal{D}+\Omega}$;

(c′) $\pm\nu_i\varepsilon_{\mathcal{D}+\Omega},\ \pm\nu_j$.

It is clear that each equivalence class may be associated to a unique frame (see Section 9) by taking differences. Since any two frames are congruent under G, we see that in the stabilizer of any one of these frames, a Sylow 11-normalizer acts transitively on the other two classes (10.4.4). It follows that the action of $N_G(\langle\alpha\rangle)$ on this set of 3 classes is Σ_3.

Since the $\mathbb{Q}$-span of any class is two dimensional and two of them are linearly independent, an element g of order 3 in $N_G(\langle\alpha\rangle)$ is fixed point free on L. Note that $g \in C_G(\alpha)$. Now, suppose that g has nonzero fixed point sublattice, R. Then, R is $N_G(\langle\alpha\rangle)$-invariant. Since $R \cap L = 0$, α acts fixed point freely on R and so the minimum polynomial for α on R is $(t^{11}-1)/(t-1)$, whence $rank(R) = 10$ or 20.

We observe that an element of order 3 in the standard frame stabilizer N_{24} is conjugate to an element of M_{24} with cycle shape $1^6 3^6$ or 3^8, and so has fixed point subllattice of dimension 12 or 8. Therefore, g does not stabilize a frame. Assuming $dim\, R \geq 10$, we let $2\Lambda \leq I \leq R$, so that $I/2\Lambda$ is totally singular; then all cosets of I have type 2. This means that if x and y are independent vectors of type 2 in I, $x.y = \pm 2$. So, take independent vectors x, y, z of type 2 in I so that $x+y$ has type 2. Then, $x.z$, $y.z$ and $(x+y).z$ are all $2 (mod\, 4)$, a contradiction. So, $dim\, I \leq 2$. This contradicts $dim\, R \geq 10$, and we have proved (iii). We leave (ii) as an exercise. □

(10.5) Theorem. *The group* .223 *is simple of order* $2^7 3^6 5^3 7.11$.

Proof. We use the faithful permutation representation of degree 275 of (10.3). Recall that $M_{22} \leq X := .223$; its index is $2025 = 3^4 5^2$.

Let K be a minimal normal subgroup of X.

Case 1. $K \cap M_{22} = 1$. Then $|K|$ divides $3^4 5^2$, whence K is solvable, by Burnside's famous $p^a q^b$ theorem [Go68] [Hu], and so is a p-group, for $p = 3$ or 5. Since X is a transitive permutation group of degree $275 = 5^2 11$, the orbits of K have the same length, whence $p = 5$. Since the simple group M_{22} acts by conjugation on $K \cong 5$ or 5^2, we conclude that $KM_{22} = K \times M_{22}$ and deduce a contradiction from (10.4.i).

Case 2. $K \cap M_{22} \neq 1$, whence simplicity of M_{22} implies that $M_{22} \leq K$. Let $P \in Syl_{11}(M_{22})$. The Frattini argument implies that $N_X(P)$ covers X/K, a group of order $3^a 5^b$, $a \leq 4$, $b \leq 2$. Since $N_{M_{22}}(P) \cong 11{:}5$, if $b \geq 1$, we contradict (10.4). So, $b = 0$ and a similar argument shows that $a \leq 1$. We are done if we show $a = 0$.

We assume a = 1. Let $m := |N_K(P)|$, $n := |K : N_K(P)|$. Then, Sylow's theorem for the prime 11 implies that $n \equiv 1 (mod\, 11)$. Since $m = 55d$, where d divides $3^3 5^2$, we conclude that $n = 2^7 3^6 5^2 7/d \equiv 1 (mod\, 11)$, or $d \equiv 1 (mod\, 11)$ and so $d = 1$ or 45. By (10.4.ii), $d \neq 45$ and so $d = 1$. We conclude that $|N_K(P)| = 55$ and $|N_X(P)| = 3.5.11$. However, (10.4.iii) implies that a group of order 33 in G has 0 fixed point sublattice in Λ, a contradiction. Therefore, $a = 0$, $K = X$ and we are done. □

(10.6) Definition. The *simple group McL of Jack E. McLaughlin* is a rank 3 permutation group (10A.1) on a graph of 275 vertices with point stabilizer isomorphic to $PSU(4,3)$.

(10.7) Remarks. The above graph has full automorphism group of the form $McL{:}2$. A simple group with an involution whose centralizer is the double cover $2\cdot Alt_8$ (Appendix 10B) is isomorphic to McL, and conversely. An involution centralizer for .223 will be discussed in (10.37). It will follow that $.223 \cong McL$, so we could define McL as the group .223.

There is a graph on 275 points of the 276 in (10.6), which we now describe. Note that any two of these vectors which do not sum to $-x$ (in the notation of (10.2.1)) has inner product 2 or 1. By double transitivity (10.3), these are the only possible inner products and so we characterize the two orbits of G_x by inner products. If we fix two vectors u, v in one orbit, the stabilizer is a subgroup $G_{x,u,v}$ of index 275 in $G_{x,u}$ and of order $2^8 3^4 5.7$. Since $274 = 2.137$, G_x is not triply transitive on the 276 points. It turns out that the rank is 3, $G_{x,u} \cong PSU(4,3).2$ and that, depending on choice of $v, G_{x,u,v} \cong PSL(3,4).2$ or $3^4{:}\,Alt_6.2$; the suborbits have lengths 1, 162 and 112. We therefore get a rank 3 graph, though additional work would be required to see that this graph is isomorphic to the original McLaughlin graph. See [McL69] for an exposition.

Besides $PSU(4,3)$, McL contains $PSU(3,5)$. We shall not attempt identifications of such subgroups by "elementary means" since a fair amount of work with classical groups and finite geometries would be needed. Note that, if we did know that McL contains the simple group $PSU(4,3)$, which would have index prime to 3, we get an easier proof of (10.5) since Case 2 would immediately lead to $a = 0$ and so X/K has order 5 or 5^2.

(10.8) Exercise. Let $x, y \in \Lambda_2$. Then, $x.y \in \{-4,-2,-1,0,1,2,4\}$. For k in this set, define $A_k(x) := \{y \in \Lambda_2 | x.y = k\}$.

(10.8.1) Prove that $Co_2 := G_x$ is transitive on all these sets and get their cardinalities. Deduce corresponding transitivity statements on triangles.

(10.8.2) Prove appropriate primitivity statements. Notice that a triangle stablizer fails to be a maximal subgroup if there is an element in Co_2 which fixes one leg and interchanges the other two.

(10.9) Theorem. *Co_2 is simple of order* $2^{18}3^6 5^3 7\cdot 11\cdot 23$.

Proof. This is an exercise. Since we have a Sylow group of order 23 and a permutation representation of degree 2300, we can play the by-now familiar game. □

(10.11) Remark. We now prepare to analyze conjugacy classes of involutions in Co_0 and Co_1 and prove (10.15). See Section 2 for background in group extension theory. Here, we shall need to study complements to an elementary abelian 2-group in an extension by a cyclic 2-group. The Jordan canonical form is the essential invariant here (2.8.2); cohomology vanishes in positive degree iff the Jordan canonical form is a sum of indecomposable blocks of maximum possible degree. We write $J_a + J_b + \ldots$ for a direct sum of Jordan blocks of degrees a, b, ... with 1 on the diagonal and superdiagonal.

(10.12) Lemma. *Let t be an involution of M_{24} with cycle shape 2^{12}. Then:*

(10.12.1) *t inverts an element of order 11 in M_{24};*

(10.12.2) *$\mathcal{G}$ and $\mathcal{G}^*$ are free modules for $\mathbb{F}_2\langle t\rangle$;*

(10.12.3) *t fixes 32 dodecads, 15 octads and 15 sixteen sets in $\mathcal{G}$.*

(10.12.4) *$C_{M_{24}}(t)$ acts transitively on each of the three types of Golay sets fixed by t;*

(10.12.5) *If $\mathcal{D}$ is a dodecad fixed by t and M_{12} its stabilizer in M_{24}, there is one M_{12}-class of involutions in $M_{12}t$. We have $C_{M_{12}}(t) \cong 2 \times \Sigma_5$.*

(10.12.6) *If $\mathcal{O}$ is an octad fixed by t, t acts with Jordan canonical form $J_1 + J_1 + J_2$ on $O_2(G(\mathcal{O}))$; the other $G(\mathcal{O})$-class of involutions which act with Jordan canonical form $J_2 + J_1 + J_1$ are in the M_{24} class of elements with cycle shape $1^8 2^8$. The $G(\mathcal{O})$-class of involutions which act with Jordan canonical form $J_2 + J_2$ have cycle shape $1^8 2^8$.*

Proof. (1) Sylow theory for the prime 11 implies that a Sylow 11-normalizer has even order. Since 11 does not divide the order of an involution centralizer in M_{24}, there is a subgroup $\langle x, u\rangle \cong Dih_{11}$, with $|x| = 11$. Since x lives in a natural M_{12}-subgroup, it has cycle structure $1^2 11^2$. Its M_{12}-normalizer has shape 11:5 (6.19.ii), whence u interchanges the two orbits of length 11 and so fixes at most 2 points of Ω and so, by the classification of involutions in M_{24} (6.41), u is conjugate to t.

(2) The 10-dimensional subspace $[\mathcal{G}, x]$ is a free $\langle t\rangle$-module, so it suffices to show that the fixed point subspace of x is a free module; but this subspace is $\{\emptyset, \Omega, D, E\}$, where D and E form a pair of complementary dodecads; t must interchange them (which implies the freeness), or else t would fix the unique point of each fixed by x.

(3) By taking u as $(11\omega\omega\bar{\omega}\bar{\omega})^{\pi_1}$ (see (6.41)), it is clear that we get 15 octads of the form $K_i + K_j$ and the complementing sixteen-sets. The dodecad consisting of one orbit of u from each of the six columns is invariant and so are its 32 transforms under $\mathcal{H}^{\pi_1}$.

(4)(5) Let $C := C_{M_{24}}(t)$. From (6.41), we know that C has shape $2^6{:}\Sigma_5$. Without loss, we may assume that t acts on $\mathcal{D}$ as the element τ of (6.20), whence $C_{M_{12}}(\tau) \cong 2 \times Alt_5$, by (6.25.5). Since this group has index 32 in C, we finish by quoting the transitivity proved above in (3).

(6) The proof of (3) shows transitivity on dodecads, so we need to treat only the case of octads. The action of t on $\mathcal{O}$ must be fixed point free. Consider the action of t on $R(\mathcal{O}) = O_2(G(\mathcal{O}))$.

We claim that a coset of $R(\mathcal{O})$ corresponding to involutions of $G(\mathcal{O})$ acting with Jordan canonical form J_2+J_2 has no conjugates of t. Such a coset consists of a single $R(\mathcal{O})$-conjugacy class of involutions. Such an involution has centralizer in $G(\mathcal{O})$ of shape $2^2{:}\left[2^2 \times Alt_4\right]{:}2$, which has order $2^7 3$; if the involution were t, this subgroup would be a stabilizer in a C-orbit of length $2^2 5 = 20 > 15$, contradiction.

Since t has the Jordan canonical form $J_2+J_1+J_1$, there are exactly two $R(\mathcal{O})$-classes of involutions represented in the coset $R(\mathcal{O})t$. We claim that elements of the other class have shape $1^8 2^8$. This is clear since there are six examples of such a u in (6.10.2).

We now have that the set of M_{24}-conjugates of t lying in $G(\mathcal{O})$ forms a single $G(\mathcal{O})$-conjugacy class. By considering the action of M_{24} on pairs $(t', \mathcal{O}')$, where t' is a conjugate of t stabilizing the octad $\mathcal{O}'$, we deduce from transitivity on octads that C has one orbit on the 15 octads fixed by t. □

(10.13) Lemma. *Let t be an involution in the natural M_{24} in N_{24}. Then t is conjugate to an ε_S.*

Proof. It suffices to show that t stabilizes a frame diagonally. From (6.41), we know that t stabilizes all parts of some sextet, Ξ. Then, t diagonally stabilizes the frame whose members are vectors of the form $\sum_{i \in T} c_i \alpha_i$, $T \in \Xi$, $c_i = \pm 1$ with evenly many minusses. □

(10.14) Lemma. *Let $tu = ut \in N_{24}$ be an involution, where t is an involution of M_{24} and $u = \varepsilon_S \in O_2(N_{24})$. Then tu is is conjugate to some $\varepsilon_R \in O_2(N_{24})$. If t has cycle shape 2^{12}, then R is a dodecad and if t has cycle shape $1^8 2^8$, $|R| = 8$, 12 or 16 and all these possibilities occur.*

Proof. It suffices to find a frame with respect to which tu acts diagonally.

We may write our involution as tu, where $tu = ut$ and u is a nonidentity element of $O_2(N_{24})$ and $t \in M_{24}$. If t has cycle type $2^{12}, tu$ is conjugate to t since $O_2(N_{24}) \cong \mathcal{G}$ is a free $\mathbb{F}_2\langle t\rangle$-module; see (2.8.2). Now use (10.13). We may therefore assume that t has cycle type $1^8 2^8$. Since t and u commute, $u = \varepsilon_S$ satisfies $S = S^t$ and so that $S \cap C = \emptyset$ or C for every cycle C of t. Let $\mathcal{O}$ be the octad fixed pointwise by t and let $U := S \cap \mathcal{O}$. By (10.12), we may assume that $S \neq \emptyset$.

We argue that $|U| \equiv 0 (mod 4)$. Since G is simple, $det\, u = 1$ and so U must be an even set. We assume that U is a 2-set or a 6-set. But then the action of t on the Golay set S has oddly many cycles, a contradiction to the fact that the stablizer of a Golay set is a perfect group of the form $2^4{:}GL(4,2), M_{12}$ or M_{24}. An alternate argument is to notice that the eigenvalue -1 has multiplicity $2 (mod 4)$, whence tu lifts to an element of order 4 in the double cover of $Co_0 \leq SO(24, \mathbb{R})$ (2.28). This contradicts the fact that the Schur multiplier of Co_0 is trivial; see (2.14).

Suppose that $|U| = 0$ or 8. Then, S is an octad or a 16-set. Under $G(\mathcal{O})$, the 30 octads disjoint from $\mathcal{O}$ form an orbit (6.9), so we may assume that $\mathcal{O} = K_{12}$ and $S = K_{34}$ or K_{1234}.

If U is a four-set, we recall that there are respectively 4, 24 and 4 Golay sets of cardinality 8, 12 and 16 which meet $\mathcal{O}$ in U. Let $C := C_{M_{24}}(t) \leq G(\mathcal{O})$. Then, $C \cong 2^4{:}AGL(3,2)$ and C acts 3-transitively as $AGL(3,2)$ on $\mathcal{O}$. We may assume that $t = r_{001111}$ (6.6), whence any $\langle t \rangle$ invariant subset of $\mathcal{O} + \Omega$ has even parity, so S has even parity. By 3-transitivity of $AGL(3,2)$, we may assume that U contains 3 points of K_2; by evenness, $U = K_2$.

Now use the following frame: $\{\alpha_{K_i}\varepsilon_A \mid A \in \mathcal{PE}(K_i), i = 1, \ldots, 6\}$. □

(10.15) Theorem. (i) *An involution in Co_0 is conjugate to an element of $O_2(N_{24})$, hence to one of ε_S, where S is an octad or its complement, a dodecad or the universe.*

(ii) *The respective centralizers are groups of the form*

(10.15.1) $2^{1+8}.W'_{E_8} \cong \left[2^{1+8} \times 2\right] \cdot \Omega^+(8,2)$, *for S an octad or 16-set;*

(10.15.2) $2^{12}{:}M_{12}$ *for S a dodecad;*

(10.15.3) Co_0, *for $S = \Omega$.*

(iii) *The group $2^{12}M_{12}$ is contained in a unique conjugate of N_{24}.*

Proof. By (10.15), it remains to determine the structures of the centralizers of the involutions $t = \varepsilon_S$.

Case 1. S is an octad or a 16-set. Without loss, S is a sixteen-set. Let $\mathcal{O} = S + \Omega$. The set of vectors of type 2 which are fixed by S is $X \cup Y$, where $X := \{\pm\alpha_i \pm \alpha_j \mid i \neq j \text{ in } \mathcal{O}\}$ and $Y := \left\{-\frac{1}{2}\alpha_{\mathcal{O}}\varepsilon_T \mid T \in \mathcal{G}\right\}$, a set of $112+128 = 240$ vectors, which span a sublattice isometric to $\sqrt{2}L_{E_8}$. Thus, we have a homomorphism π from $C := C_G(t) \to W_{E_8}$. Notice that the image of $C_N(t) \cong 2^{12}{:}2^4{:}GL(4,2)$ has image of the shape $2^7{:}GL(4,2)$, which is a parabolic subgroup of $W'_{E_8} \cong 2 \cdot \Omega^+(8,2)$; see (8.24). The kernel of π is a group $Q := \langle \varepsilon_S, R(\mathcal{O}) \mid S \in \mathcal{G}, S \cap \mathcal{O} = \emptyset \rangle$; by (6.8), $Q \cong 2_+^{1+8}$ (2.20).

Now, let E be a "partial frame," i.e., a subset of of $L \cong \sqrt{2}L_{E_8}$ corresponding to $\sqrt{2}F$, where F is as in (8.24), and denote by $E_0 = \{\pm\alpha_i \mid i \in \mathcal{O}\}$ the intersection of the standard frame with L. We claim that E is conjugate to E_0 by an element of $N(Q)$. If an 8-set in Ω is fixed pointwise by an involution in M_{24}, it is an octad and is fixed pointwise by a conjugate of $R(\mathcal{O})$. Letting $g \in G$ conjugate the frame containing E to the standard one, we see that Q^g is not in $O_2(N_{24})$, so the previous sentence implies that there is $h \in M_{24}$ such that $E^{gh} = E_0$. Since Q^{gh} centralizes E_0 and its rational span, $Q^{gh} \leq Q$ and so $Q^{gh} = Q$. The claim follows. Since there are 135 such partial frames (8.24), we use the claim and the previous paragraph to conclude that $|N(Q)/Q| = 2^{12}3^5 5^2 7$ and so $Im(\pi)$ is the commutator subgroup of W_{E_8}.

Case 2. S is a dodecad. Then, the vectors of type 2 in the fixed points of t look like $X \cup Y$, where $X := \{\pm\alpha_i \pm \alpha_j \mid i \neq j \text{ in } S+\Omega\}$ and $Y := \left\{-\frac{1}{2}\alpha_S\varepsilon_T \mid T \in \mathcal{G}\right\}$. The sublattice $C_\Lambda(t)$ is isometric to $\sqrt{2}HS_{12}$ (see (8.18)) and so has automorphism group $2^{11}{:}\Sigma_{12} \cong W_{D_{12}}$. The natural map π of $C := C_G(t)$ to $Aut(HS_{12})$ has the property that $C_N(t)^\pi$ has the form $2^{11} : M_{12}$ and $Ker(\pi) = \langle t \rangle$ (reason: $C_{C(t)}(X \cup Y)$ stabilizes $8\alpha_k, k \in S$, so lies in N_{24}, the standard frame group). Thus, as $C(t)$ acts on $C_\Lambda(t)$, it normalizes $O_2(N_{24})$ and so is in the maximal subgroup $N_{24} = N_G(O_2(N_{24}))$, whence $C = C_N(t)$ .0

(10.16) Corollary. *$O_2(N_{24})$ is weakly closed in N_{24} with respect to G. (Compare with* (6.12.4).*)*

Proof. Let W be a conjugate of $U := O_2(N_{24})$ in $N := N_{24}$, but $U \neq W$. Since $U = C(U), C_U(W) = U \cap W = C_W(U)$. Let $2^a := |U \cap W|$.

Suppose that $U \cap W$ does not contain $\varepsilon_\mathcal{D}$, for any dodecad $\mathcal{D}$. By (6.43), $a \leq 5$. By (6.45), $12 - a \leq 6$, so $a = 6$ and by (6.43), there is an octad $\mathcal{O}$ so that $U \cap W = \langle \varepsilon_\mathcal{O} \rangle \times \{\varepsilon_S \mid S \in \mathcal{G}, S \cap \mathcal{O} = \emptyset\}$. Then $W/W \cap U$ embeds in the subgroup of M_{24} which fixes all Golay sets which avoid $\mathcal{O}$. Taking intersections, we deduce that this subgroup fixes $\Omega + \mathcal{O}$ pointwise, so is the identity, a contradiction.

We conclude that $U \cap \mathcal{D}$ contains some $\varepsilon_\mathcal{D}$. Thus, $W/W \cap U$ embeds in M_{12}, which has 2-rank 3 (6.46). Therefore, $a \geq 9$. For any involution $t \in M_{24}$, $dim\, C_\mathcal{G}(t) \leq 8$ (6.47), (10.12.2). Consequently, since $W \neq U$, $a \leq 8$, a contradiction. □

(10.17) Corollary. *$\{F \mid F$ is a frame fixed by $O_2(N_{24})\}$ is just the standard frame.*

In the following result, we identify the "projective involutions" and get partial information on the centralizers.

(10.18) Theorem. *Let $\mathcal{P}$ be the set $\{g \in G \mid g^2 = -1\}$. Then $\mathcal{P}$ is a conjugacy class in G whose centralizers C have order $2^{15}3^3 5^2 7 \cdot 13$. Also, the lattice $[\Lambda, g]$ satisfies: $2\Lambda < [\Lambda, g] < \Lambda$, $[\Lambda, g] \cong \sqrt{2}\Lambda$, $[\Lambda, g]/2\Lambda$ is a maximal totally singular subspace of $\Lambda/2\Lambda$ and C operates transitively on $\left([\Lambda, g]/2\Lambda\right)^\#$. The set $N_{24} \cap \mathcal{P}$ is a conjugacy class in N_{24} and the centralizer in N_{24} of such an element has the form $2^{6+6}{:}\Sigma_5$.*

Proof. Let $N := N_{24}$ and let $\mathcal{F}$ be the set of frames in Λ; a stabilizer of a frame is a conjugate of N. Define

(10.18.1)
$$\mathcal{Q} := \{(F, g) \in \mathcal{F} \times \mathcal{P} \mid F = F^g\} .$$

Since N contains a Sylow 2-group of G, we may take a representative element of $\mathcal{P}$ from N. Such an element must have the form $\varepsilon_S t$, where t is a permutation of order 2 such that $S + S^t = \Omega$, whence S is a dodecad.

Since $dim\, C_\mathcal{G}(t) = 6$, there are 64 such dodecads (the inverse image of Ω under the linear map commutation with t); let $\mathcal{X}$ be the set of such dodecads. The stabilizer of $\mathcal{D} \in \mathcal{X}$ in M_{24} is M_{12} and the centralizer in M_{12} of an outer automorphism of order 2 is isomorphic to $2 \times Alt_5$; see (10.12.4 and 5). From (6.41), we know that $C_{M_{24}}(t) \cong 2^6{:}\Sigma_5$, whence $C_{M_{24}}(t)$ has a single orbit of length 64 on

$\mathcal{X}$. Therefore, in N, there are $|M_{24}|/2^3 3.5 = 2^7 3^2 7.11.23 = 1771.18.64 = 2040192$ elements of $\mathcal{P}$.

It follows that all elements of $\mathcal{P}$ of the form $\varepsilon_S t$ form a single conjugacy class under $C_{N_{24}}(t)$ (even under $C_{M_{24}}(t)$). Therefore, transitivity of G on frames implies transitivity of G on $\mathcal{Q}$. Transitivity of G on $\mathcal{P}$ follows from the definition of $\mathcal{Q}$.

We know from the transitivity of G on $\mathcal{Q}$ that $C(g)$ is transitive on the frames stabilized by g. Let n be the number of such frames. We claim that $n = 2^{12} - 1 = 4095 = 3^2 5.7.13$ and that g acts as an involution on $\Lambda/2\Lambda$ with 12 Jordan blocks of degree 2. Since $g^2 = -1, [\Lambda, g] \geq [\Lambda, g^2] = 2\Lambda$. The stabilizer in G of a coset of type 2 or 3 has the form $2 \times X$, for a simple group X. Therefore, $g^2 = -1$ implies that M has nontrivial cosets of type 4 only. Since $M := [\Lambda, g]/2\Lambda$ is totally isotropic, it has dimension at most 12; since it has at least 4095 cosets of type 4, $dim\, M = 12$ and $C(g)$ operates transitively on the set of its nonzero vectors. This proves the claim. From $|G{:}C(g)|3^2 5.7.13 = |G{:}C(g)|n = |\mathcal{Q}| = |G{:}N|2^7 3^2 7.11.23 = 2^7 3^8 5^3 7^2 11.13.23$, we deduce that $|C(g)| = 2^{15} 3^3 5^2 7 \cdot 13$.

Transitivity of G on $\mathcal{Q}$ also implies that $N \cap \mathcal{P}$ is a conjugacy class. If $g \in N \cap \mathcal{P}, |C_N(g)| = 2^{15} 3.5$, so it remains to determine the shape of $C_N(g)$. Letting $g = \varepsilon_S t$, as above, we put g in the subgroup $O_2(N)H$, where H is a sextet stabilizer in M_{24} and $t \in O_2(H)$. From (10.12.1), we see that $C_{O_2(N)}(t) \cong 2^6$ and from (10.12.4) we see that $C_H(\langle \varepsilon_S, t\rangle) \cong 2 \times Alt_5$; if $x \in C_H(t) \cong 2^6{:}\Sigma_5$ is an involution corresponding to a transvection on the Σ_5-quotient, then there is an element $y \in O_2(H)$ satisfying $S^x = S^y$. Then, xy centralizes g and we deduce that $C_N(g) \cong 2^{6+6}{:}\Sigma_5$. □

(10.19) Remark. It turns out that C has the form $8 \cdot G_2(4) = [4 \cdot G_2(4)] : 2$, i.e., $C' = C''$ is the covering group of $G_2(4), Z(C) \cong 4$ and $C/Z(C) \cong Aut(G_2(4))$. Also, $N_G(C') = [S \circ C'] \langle t \rangle$, where $S \cong SL(2,3)$ and $|t| = 2$. We can continue the argument of (10.18) a little further, but to identify $C'/Z(C)$ as $G_2(4)$, we need to do something like determine suitable generators and relations (2.27.4), or find the centralizer of an involution in a long root group [Thomas]. The double cover of $G_2(4)$ is analyzed in [Gr72b].

Similar considerations apply when trying to identify many of the groups in the table of Appendix 10C. It seems reasonable that the ones of Lie type can be identified by locating suitable generators in the sense of (2.27).

(10.20) Lemma. (i) *Let $g \in G$ have order* 23. *The theta series of the fixed point sublattice of g begins* $1 + 2q^2 + 2q^3 + 2q^4 + \ldots$.

(ii) *Let L be a sublattice of Λ containing a triangle of type* 233. *Then L is a direct summand of Λ and L has exactly four elements of type* 3.

Proof. An exercise; see the proof of (9.15). □

(10.21) Theorem. *There is one G-orbit on triangles of type* 233. *Let H be the stabilizer of one. Let $H < K \cong Co_3$, the group stabilizing a leg x of type* 3.

(i) *There is one G-conjugate of H in K.*

(ii) *On the* 276 *triangles of type* 223 *containing* x (10.5), *the orbits of H have length* 100 *and* 176.

(iii) *The action of H on the orbit of length* 100 *is rank* 3 *with point stabilizer a natural* M_{22} *subgroup of G.*

(iv) *The action on the orbit of length* 176 *is doubly transitive with stabilizer of order* $|PSU(3,5){:}2| = 2^5 3^2 5^3 7$. *(The stabilizer is isomorphic to* $PSU(3,5)$, *extended by a group of field automorphisms of order* 2*; we do not prove this; see* (10.19).*) In fact, the natural* M_{22} *subgroup stabilizing a triangle in the length* 100 *orbit is transitive on the length* 176 *orbit.*

(v) *H is simple of order* $2^9 3^2 5^3 7.11$.

(vi) *The group* H^*, *defined as the stabilizer of the set of triangle legs, contains H with index* 2.

(vi.a) *Elements of* $H^* \backslash H$ *induce outer automorphisms on H and extend the action on the degree* 100 *representation; in* H^* *a point stabilizer is isomorphic to* $M_{22}{:}2$, *the stabilizer in* M_{24} *of a* 2*-set.*

(vi.b) *The action of H on the degree* 176 *representation does not extend to* H^**; thus, H has two inequivalent degree* 176 *doubly transitive representations.*

Proof. Transitivity is an exercise (10.4.iii). Transitivity of G on Λ_3 implies (i). Without loss, we take $x = \frac{1}{4}\left(5\alpha_i + \sum_{j \neq i} \alpha_j\right)$ and consider triangles of type 223 containing x as a leg; see (10.4.1). Let U be the triangle containing x and $z = -\alpha_i + \alpha_j$; so, U has type 233; we take K as the stabilizer of U. Then $K \cap N$ is a natural M_{22} subgroup with orbits of lengths 1,22 and 77 on the trangles of type 223 containing $-\nu_j$ as a leg. The ones containing $\nu_i \varepsilon_S$, where S is a 16-set not containing i, fall into a single M_{22} orbit of length 176 (see (6.10) for the properties of an octad stablizer). Since K fixes no triangle of type 223 (quick proof: neither of $|.223|$, $|.233|$ divides the other), K has no orbit of length 1. Also, K cannot have the orbits of length 1,22 or their union as an orbit, since K is not contained in a frame stabilizer (9.10,14). By Lagrange, K is not transitive on the set of $276 = 23.12$ triangles. Thus, there is an orbit of K of length dividing 1 plus a nonempty, proper subsum of $\{22, 77, 176\}$; Lagrange's theorem permits only $1 + 22 + 77 = 100$. This proves (ii) and (iii). □

A different proof can be given by using constraints on inner products. Show that:

(10.21.1) the vectors of the set $A_{-3}(x)$ (10.4.1) fall into the following classes, according to their inner products with $z = -\alpha_i + \alpha_j$ (note that $(z, x) = -2$; S is a 16 set and $\mathcal{O} = S + \Omega$ is an octad):

77 vectors $\nu_{i,S}$ $(j \notin S)$	inner product 2
176 vectors $\nu_{i,S}$ $(j \in S)$	inner product 1
1 vector $-\nu_j$	inner product 2
22 vectors $-\nu_k$ $(k \neq i, j)$	inner product 0
1 vector $-\alpha_i - \alpha_j$	inner product 0
22 vectors $-\alpha_i - \alpha_k$ $(k \neq i, j)$	inner product 2
176 vectors $\frac{1}{2}\alpha_{\mathcal{O}}$ $(i \in \mathcal{O}, j \notin \mathcal{O})$	inner product 1
77 vectors $\frac{1}{2}\alpha_{\mathcal{O}}$ $(i \in \mathcal{O}, j \in \mathcal{O})$	inner product 0.

To prove (iv), we note that H is transitive on the 176 objects because a natural M_{22} subgroup is transitive. If H_0 is a point stabilizer in H, $H_0 \cap M_{22}$ is a natural Alt_7 subgroup, with orbit structure $1 + 7 + 1 + 15$ on Ω. Its orbit structure on the 176 objects is $1 + 70 + 105$, due to the following facts: M_{24} is transitive on the $6800640 = 2^8 3.5.7.11.23$ ordered quadruples $(i, j, \mathcal{O}, \mathcal{O}')$, where i and j are distinct points, $\mathcal{O}$ and $\mathcal{O}'$ are octads such that $i \in \mathcal{O} \cap \mathcal{O}'$ and $j \notin \mathcal{O} \cup \mathcal{O}'$. The action is rank 3, distinguished by the property that when $i = j$, $\mathcal{O} \cap \mathcal{O}'$ is a 2-set or a 4-set. A count of such configurations shows that the intersection 2 case gives an orbit of Alt_7 of length 70 and the other leads to an orbit of length 105; the verification is left as an exercise with the material of Sections 5 and 6. Now, if H were not doubly transitive, it would be of rank 3 and H_0 would fix each suborbit and so fix the sum of the vectors $\nu_i \varepsilon_S$ in the length 70 suborbit (S ranges over the 16-sets which avoid i but contain j). A look at the Leech triangle (5.1) shows that this sum is $\frac{1}{4}\left[-210\alpha_i - 70\alpha_j - 64\sum_{k \neq i,j} \alpha_k\right]$. Since H_0 also fixes x and z, it follows that H_0 fixes α_i and α_j and so stabilizes the standard frame. Thus, H_0 is in a natural $2^{10}M_{22}$ subgroup of N_{24}. However, we know that $|H_0| = 2^5 3^2 5^3 7$, so we have a contradiction to Lagrange's theorem. Double transitivity follows.

(v) Let J be a normal subgroup of H; double transitivity implies primitivity, whence J is transitive on the $176 = 2^4 11$ points (1.2) and so contains a Sylow 11-subgroup of G. Since M_{22} is simple (6.42.4), J contains the above natural M_{22} subgroup, whence the index of J divides 100. But now, primitivity (iii) on the rank 100 orbit of H implies that J is transitive there. Since this M_{22} is a point stabilizer, $J = H$.

(vi) This is easy to see if we take the type 223 triangle

$$(10.21.1) \quad x_i := \tfrac{1}{4}\left(5\alpha_i + \textstyle\sum_{k \neq i} \alpha_k\right),\ x_j := -\tfrac{1}{4}\left(5\alpha_i + \textstyle\sum_{k \neq j} \alpha_k\right),\ z = -\alpha_i + \alpha_j;$$

just use an appropriate element of N_{24}, say $g = -p$, where $p \in M_{24}$ interchanges i and j.

(vi.a) Let X be the set of 100 vectors with inner product 0 in (10.21.1). Then $X^g = -X$. It follows that p stabilizes X and induces on H the same automorphism as g. Obviously, p normalizes the natural M_{22} in H.

(vi.b) Let Y be the set of 352 vectors in (10.21.1) with inner product 1. Then $Y^g \neq \pm Y$, so we cannot proceed as in (vi.b). Let $H_0 \cap M_{22} \cong Alt_7$ be as in the proof of (iv). We may choose p from the proof of (vi) to satisfy $\mathcal{O} \cap \mathcal{O}^p = \emptyset$

and even to be an involution in the factor $L_2 \cong \Sigma_3$ in the stabilizer of the trio $\{\mathcal{O}, \mathcal{O}^p, \Omega + \mathcal{O} + \mathcal{O}^p\}$ (6.10) which centralizes $L_{3,1}$ (6.11). It follows from the action of p on the three octads that $[H_0 \cap M_{22}] \cap [H_0 \cap M_{22}]^p$ has orbit structure $1 + 7 + 1 + 7 + 8$ on Ω. Since $L_{3,1} \leq [H_0 \cap M_{22}] \cap [H_0 \cap M_{22}]^p$, and $L_{3,1}$ is a maximal subgroup of $H_0 \cap M_{22}$, $L_{3,1} = [H_0 \cap M_{22}] \cap [H_0 \cap M_{22}]^p$. Thus, on Y^g, $H_0 \cap M_{22}$ has an orbit of length 15. Since this is not so on Y, the two degree actions of H we find in Y and in Y^g are inequivalent.

(10.22) Remark. If U is a point stabilizer in this representation of H and $x \in H^* \backslash H$, then $U \cap U^x \cong \Sigma_7$ or $5^{1+2}{:}SD_{16}$ and both possibilities occur. Their indices in U are 50 and 126.

(10.23) Definition. The *group HiS of Donald G. Higman and Charles Sims* is a rank 3 permutation group on 100 points in which a point stabilizer is isomorphic to M_{22}.

(10.24) Remarks. (i) For characterizations of this rank 3 graph, see [Wales]. The group HiS is simple and the full automorphism group of this graph is a group $HiS{:}2$.

(ii) The Higman-Sims group has involution with centralizer of the form $4 \circ 2^{1+4}.\Sigma_5$ and another involution with centralizer $2^2 \times Aut(Alt_6)$. The analysis of centralizers of 2-central involutions in .233 is left as a hard exercise. See the discussion for involution centralizers in McL (10.36).

We now move on to study certain local subgroups of G. Our classification of involution centralizers will be an important tool. We start with an analysis of the conjugacy classes of elements of order 3.

(10.25) Notation. Recall the notations $\varepsilon_{\mathcal{D}}$, $\varepsilon_{\mathcal{O}}$, $\varepsilon_{\mathcal{O}+\Omega}$ from (9.1); here, $\mathcal{O}$ is a fixed octad. The term *spectrum* means the spectrum of an endomorphism on 24-space containing Λ, unless stated otherwise.

(10.26) Lemma. (i) *There are two classes of elements of order* 3 *in* $C(\varepsilon_{\mathcal{D}}) \cong 2^{12}{:}M_{12}$. *They are conjugate to elements of the standard* M_{24} *subgroup with respective cycle shapes* $1^6 3^6$, 3^8 *and spectra* $1^{12}\omega^6\bar{\omega}^6$, $1^8\omega^8\bar{\omega}^8$. (ii) *For the latter class, the centralizers in G have shape* $3 \times 2^{\boldsymbol{\cdot}}Alt_9$; *the unique classes of noncentral involutions of this centralizer are conjugate to* ε_D.

Proof. (i) This is clear from (10.16.2) and the facts that M_{12} and M_{24} have two classes of elements of order 3 (7.35).

(ii) Say $v \in M_{24}$ has cycle shape 3^8. Then v fixes exactly 16 Golay sets: $\emptyset$, Ω and 14 dodecads. In M_{24}, the centralizer of v is $\langle v \rangle \times L$, where $L \cong L_2(7)$ acts with orbits $8 + 8 + 8$ (6.42.3), hence acts transitively on the 14 dodecads above. See (7.33).

In L, any involution acts on $U := \{\varepsilon_A \mid A \text{ is a Golay set fixed by } v\} \cong 2^4$ with Jordan canonical form $2J_2$ (this is a property of the essentially unique indecomposable L-module of dimension 4 with trivial submodule; another way to see this is that if $x \in L$ has order $k \not\equiv 0(mod\, 3)$ and x is fixed point free on Ω, the orbits of $\langle x, v\rangle = \langle xv\rangle$ on Ω have length $3k$, so if $k = 2, 4$, the number of v-fixed Golay sets fixed by x is $8/k$).

From the above, we conclude that every involution in $UL \backslash U$ has cycle shape 2^{12} and so is G-conjugate to $\varepsilon_{\mathcal{D}}$. It follows that UL contains $C_{C(v)}(t)$, for any $t \in U \backslash \langle \varepsilon_\Omega \rangle$ by (10.16.2). Also, (10.16.2) and the proof of (i) imply that there is one G-class of pairs (x, y), where x is conjugate to t, y is conjugate to v and $xy = yx$. Therefore, modulo $\langle \varepsilon_\Omega\rangle, t$ is not weakly closed in its centralizer modulo $\langle \varepsilon_\Omega \rangle$, which has shape $2^3{:}\Sigma_4$. We now quote a classification (Appendix 10B) which says that $C(v)/\langle v\rangle \cong Alt_8$ or Alt_9. Since the module U is indecomposable for L, the group $\langle v, \varepsilon_\Omega\rangle / \langle v \rangle \cong 2$ is not a direct factor of $C(v)/\langle v\rangle$, whence $C(v)/\langle v\rangle \cong 2{\cdot}Alt_8$ or $2{\cdot}Alt_9$ (2.14).

We are done once we eliminate Alt_8. Assume that Alt_8 occurs here; its Sylow 3-group is 3^2. We may assume, by Sylow's theorem for G, that v is in the subgroup $\mathcal{TG}{:}\,Aut(\mathcal{TG})$ (9.24.3). Since $C(v)/\langle v, \varepsilon_\Omega\rangle \cong Alt_8$, the action of v on $\mathcal{TG} \cong 3^6$ has at most 2 dimensions of fixed points, so the action has Jordan canonical form $2J_3$. Thus, $\mathcal{TG}$ is a free module for $\langle v\rangle$ and so the centralizer of v covers the centralizer in $\mathcal{TG}{:}\,Aut(\mathcal{TG})$ of v modulo $\mathcal{TG}$; since $|\,Aut(\mathcal{TG})\,|_3 = 3^3$ we deduce $|C_G(v)\,|_3 \geq 3^{2+2} = 3^4$, a contradiction. □

(10.27) Notation. Let $z = \varepsilon_{\mathcal{O}+\Omega}$, for an octad $\mathcal{O}$, $C := C(z)$, $Q := [O_2(C), C] \cong 2_+^{1+8}$ and let $\vartheta_k, k \in \{1,2,3,4\}$, be commuting elements of order 3 in C such that $[Q, \vartheta_k] \cong 2^{1+2k}$; then $C_Q(\vartheta_k) \cong 2^{1+2(4-k)}$. If we write ϑ_4 as a product $\alpha\beta\gamma\delta$ of commuting conjugates of ϑ_1, we get an element $\vartheta_5 := \alpha\beta\gamma\delta^{-1}$ not conjugate to ϑ_4 in C. By Witt's Theorem (2.25) for $O^+(8,2)$, modified for $\Omega^+(8,2)$, these five elements represent the conjugacy classes of elements of order 3 in C.

(10.28) Lemma (Centralizers of semisimple elements in Orthogonal Groups over $\mathbb{F}_q$). *Let V be the vector space with nondegenerate quadratic form and associated bilinear form, f. If g is a semisimple element, write $V = \oplus_{i \in I} V_i$ as a sum of homogeneous components for $\langle g\rangle$. The centralizer of g is the direct product of its centralizers of its restrictions to the V_i', where $V_i' = V_i$ if V_i is self dual and $V_i' = V_i \oplus V_j$, where V_i and V_j are dual submodules.*

If V_i is self-dual, the irreducible submodules are 2-dimensional and the centralizer is the unitary group preserving a hermitian form with base field $\mathbb{F}_q$.

If V_i is not self dual, let V_j be the homogeneous component containing the dual irreducibles. Let K be the subfield of $GL(V_i)$ generated by $\mathbb{F}_q$ and the action of g; $dim_{\mathbb{F}_q} K = d$. Then, the centralizer of g on V_i' is isomorphic to $GL(m, K)$, where $md = dim\, V_i$, and it acts faithfully on V_i and dually on V_j.

Proof. Exercise.

(10.29) Lemma. *The elements of M_{24} with cycle shape $1^6 3^6$ is G-conjugate to ϑ_2, and to one of ϑ_4 and ϑ_5, say ϑ_4 (so $\vartheta_2 \sim_G \vartheta_4$). Their common spectrum is $1^{12}\omega^6\bar{\omega}^6$.*

Proof. Let $L(\mathcal{O}) \cong GL(4,2)$ be in the M_{24} octad stabilizer. Then the subgroup $O_2(N_{24})L(\mathcal{O})$ is in C. From (10.16), we know that $O_2(N_{24}) \cap Q$ is just the kernel of the action on $C_\Lambda(z) \cong \sqrt{2} \cdot L_{E_8}$, whence $O_2(N_{24}) \cap Q = \{\varepsilon_A \mid A \in \mathcal{G}$ is disjoint from $\mathcal{O}\} \cong 2^5$, and this is a maximal elementary abelian 2-subgroup of Q. The action of $L(\mathcal{O})$ on $Q/Z(Q)$ is given by its action on the natural 4-dimensional module and its dual. Clearly then we have conjugates of ϑ_2 and ϑ_4 or ϑ_5 in $L(\mathcal{O})$. Any element of order 3 in $L(\mathcal{O})$ fixes a point of $\mathcal{O}$ hence is in the M_{24}-class with cycle shape $1^6 3^6$. □

(10.30) Lemma. *Use the notation (10.27). Let $\Lambda_\pm$ be the sublattice of vectors where z acts as ± 1. Then we have the following partial information on traces for elements of C:*

	ϑ_1	ϑ_2	ϑ_3	ϑ_4	ϑ_5
on Λ_-:	-8	4	-2	1	1
on Λ_+:	?	2	?	5	?.

Proof. The traces on Λ_- can be deduced from (2.22) and the second line for ϑ_2 and ϑ_4 are deduced from the first and the previous lemma. □

(10.31) Lemma. $C_C(\vartheta_4) \cong 3 \times 2^2 . SU(4,2)$ *and this acts as W'_{E_6} on Λ_+ and with irreducibles of dimensions 5,5 and 6 on Λ_- (these are the eigenspaces of ϑ_4).*

Proof. The first statement follows from the previous lemma. The second follows from the facts that C/Q acts faithfully on Λ_+ and that the 6-dimensional fixed point space for the action of ϑ_4 on Λ_+ admits $SU(4,2) \times 2$; furthermore, every element of this $SU(4,2)$ is a word in reflections at roots in this subspace [Carter], 2.5.5. By order considerations, the action is that of W'_{E_6}. Finally, we get the third from the character table of $SU(4,2)$ and the spectrum of ϑ_4 (10.30). □

(10.32) Corollary (Spectra of the ϑ_k). *Let $\Lambda_\pm$ be the sublattice of vectors where z acts as $\pm$ 1. Then we have the following traces for elements of C:*

	ϑ_1	ϑ_2	ϑ_3	ϑ_4	ϑ_5
on Λ_-:	-8	4	-2	1	1
on Λ_+:	-4	2	-1	5	-4
on Λ:	-12	6	-3	6	-3.

Proof. In its action on $Q/Z(Q)$, $U(4,2) = 3 \times SU(4,2)$ contains a subgroup $\langle\alpha\rangle \times \langle\beta\rangle \times \langle\gamma\rangle \times \langle\delta\rangle$ with each of $\alpha, \beta, \gamma, \delta$ conjugate to ϑ_1. We take the element $x = \alpha\beta\gamma\delta$ to be central in this copy of $SU(4,2)$. Then $y = \alpha\beta\gamma^{-1}\delta^{-1}$ is C-conjugate to x and the ratio xy^{-1} is conjugate to ϑ_2 and on Λ_+ it has spectrum $1^4\omega^2\bar{\omega}^2$ since $\langle x\rangle \neq \langle y\rangle, x$ and y commute and have spectra $1^6\omega^1\bar{\omega}^1$. Therefore, ϑ_2 has trace 2 on Λ_+ and trace 6 on Λ.

A similar trick gets the story for ϑ_3, namely the product wxy of $w = \alpha\beta\gamma\delta, x = \alpha\beta^{-1}\gamma^{-1}\delta, y = \alpha\beta\gamma^{-1}\delta^{-1}$ is conjugate to ϑ_3 and on Λ_+ it has spectrum $1^a\omega^b\bar{\omega}^b$ with $a \geq 2$ and a even. If $a \geq 4$, wxy is a product of at most 2 commuting conjugates of ϑ_1, while wxy is conjugate to ϑ_3, a contradiction. So, $a = 2$ and $b = 3$. This means that the trace of w on Λ_+ is -1.

We are left to settle the traces for ϑ_k, for $k \in \{1, 5\}$.

We now refer to the elements $\alpha, \beta, \gamma, \delta$, commuting conjugates of ϑ_1. Let $[ijkl]$ stand for $\alpha^i\beta^j\gamma^k\delta^l$. Then, $\vartheta_4 = [1111]$ and its C-conjugates in $\langle\alpha, \beta, \gamma, \delta\rangle$ are the elements $[1111], [1221], [1122]$ and $[1212]$ and their inverses. The degree 4 matrix with these vectors as rows has determinant 2, so these four elements generate $\langle\alpha, \beta, \gamma, \delta\rangle$. Each of these four elements acts with spectrum $1^6\omega\bar{\omega}$ with fixed point sublattice in Λ_+ of type $\sqrt{2}E_6$ whose orthogonal is a lattice of type $\sqrt{2}A_2$. Therefore the group $\langle\alpha, \beta, \gamma, \delta\rangle$ fixes the summands of a sublattice of type $\sqrt{2}A_2A_2A_2A_2$ in the E_8-lattice. Clearly then, $[1000]$, which is the product of these four elements, is conjugate to ϑ_1 and is represented on Λ_+ by a transformation with spectrum $\omega^4\bar{\omega}^4$. Finally, $[2111] = [1111]^2\,[1221]\,[1122]\,[1212]$ has the same spectrum as $[1000]$ and is conjugate to ϑ_5. □

(10.33) Lemma. *An element of order* 3 *in* G *commutes with an involution.*

Proof. We produce a subgroup of G where this may be seen. We make use of the construction of the Leech lattice by the ternary Golay code (9.24). This embeds $\mathcal{TG}{:}\, Aut\,(\mathcal{TG})$ in G.

We now prove that, for each element of order 3 in $\mathcal{TG}{:}\, Aut(\mathcal{TG})$ is in the centralizer of involution. We first do this for elements in $\mathcal{TG}$.

If $v \in \mathcal{TG}$ has weight 9, the stabilizer of $\langle v\rangle$ in $Aut\,(\mathcal{TG})$ has the form $2 \times 3^2\, Quat_8$. Since its derived group centralizes v, we are done.

If v has weight 6, the stabilizer of $\langle v\rangle$ in $Aut\,(\mathcal{TG})$ has the form $2.\Sigma_6 = [2 \times Alt_6]\,.2$ (this is so from (2.14) and (2.28)). Since the commutator subgroup contains involutions, we are done.

If v has weight 12, the stabilizer of $\langle v\rangle$ has index 12 and so is, by (6.25.3), of the form $2.M_{11}$; this group is isomorphic to $2 \times M_{11}$ (2.14). Since M_{11} is simple of even order, v commutes with an involution.

We now turn to elements of order 3 in $H := \mathcal{TG}{:}\, Aut(\mathcal{TG}) < G$ which are outside $\mathcal{TG}$. In $Aut\,(\mathcal{TG}) \cong 2^{\cdot}M_{12}$, there are two classes (7.33). If x represents the class with cycle shape 3^4, $\mathcal{TG}$ is a free $\langle x\rangle$-module and so the Frattini argument says that $C(x)$ covers $C_{H/O_{2,3}(H)}(x)$, which, by (7.33), is isomorphic to $3 \times Alt_4$, a group of even order. Now, let $y \in Aut\,(\mathcal{TG})$ have order 3 and cycle shape 1^33^3. The M_{12}-centralizer has the form $3^{1+2} : \Sigma_3$ and so its centralizer D in $Aut\,(\mathcal{TG})$ has the form $2.3^{1+2}{:}\,\Sigma_3$. The Jordan canonical form for y on $\mathcal{TG}$ is $J_1J_2J_3$, which means that the first cohomology group for $\langle y\rangle$ on $\mathcal{TG}$ is $\mathbb{F}_3^2$. We get that every element in the coset $\mathcal{TG}y$ commutes with an involution in $\mathcal{TG}D$ if we know that every orbit of D on the cohomology group has length not divisible by 4. By (7.37), the orbit lengths are 1, 2 and 6, so we are done. □

(10.34) Theorem. *There are exactly four classes of elements of order* 3 *in* G.

Proof. Use the notation of (10.30). By the last two Lemmas, it suffices to prove that $t := \vartheta_3$ and $u := \vartheta_5$ are conjugate in G. Let $C_C(u) = \langle u \rangle \times L$ (10.29), where $L \cong 2^2.PSU(4,2)$ (10.31). Then, on Λ_-, u has spectrum $1^6\omega^5\bar{\omega}^5$ and these eigenspaces are irreducibles for L. Also, L preserves the two 4-dimensional eigenspaces for u. The character table of L implies that $L \cong Sp(4,3) \times 2$ and, in Atlas notation, class 2A of $PSU(4,2)$ lifts to an involution in L' of trace -8 and class $2B$ lifts to an element of order 4 in L' with trace 4 on Λ. Finally, $Z(L') = \langle \varepsilon_O \rangle$.

Now consider the action of G on the set A of all pairs (x,y), of commuting elements where x is conjugate to $\varepsilon_{O+\Omega}$ and y is conjugate to ϑ_5. Our analysis of elements of order 3 implies that there are one or two orbits (as ϑ_3 or ϑ_5 are not or are conjugate in G). We now consider the action of $C(y)$ on $\mathcal{B} := \{x \mid (x,y) \in \mathcal{A}\}$. Notice that the structure of $C(zy) = \langle y \rangle \times 2 \times Sp(4,3)$ implies that z inverts $O(C(y))/\langle y \rangle$ under conjugation. Therefore, if we let z' be the element of trace -8 in L' (see above), we notice that z and z' are conjugate in G but not in $C(y)$ (since z is central modulo $O(C(y))$ and z' is not). Therefore, $C(y)$ has two orbits on $\mathcal{B}$, whence G has two orbits on A and so ϑ_3 and ϑ_5 are conjugate. □

The final step is the description of the centralizers of these elements.

(10.35) Theorem. *All elements of order 3 in G are real and there are four conjugacy classes. Representing elements have the G-centralizers as follows:*

(10.35.1) $C(\vartheta_1) \cong 6 \cdot Suz;$

(10.35.2) $C(\vartheta_2) \sim C(\vartheta_4) \cong [2 \times 3^2] \cdot PSU(4,3).2;$

(10.35.3) $C(\vartheta_3) \sim C(\vartheta_5) \cong 3^{1+4}{:}\, Sp(4,3) \times 2;$

(10.35.4) $C(g) \cong 3 \times 2 \cdot Alt_9;$ $(g \in M_{24}$ has cycle type $3^8);$

Proof. In (10.26.ii), we determined $C(g)$. As with that case, we must quote non-trivial results to make the identifications here.

For $X := C(\vartheta_3)$, we know that $C_C(\vartheta_3) \cong Sp(4,3) \times 2$ and this means that $z = \varepsilon_{O+\Omega}$ inverts $Y := O(X)/\langle \vartheta_3 \rangle$. Since z lies in the center of an extraspecial group 2_-^{1+4} in X, we have that $|Y|_p = p^d$, for $d \equiv 0 \,(mod\, 4)$ and every prime divisor d of $|Y|$. If $p \geq 5$, $|G|_p \geq p^4$ implies $p = 5$. But then $|Sp(4,3)| = 2^7 3^4 5$ implies that $|X|_5 \geq 5^5$, a contradiction. So, Y is a 3-group. Since $|G|_3 = 3^9$, we conclude that $|Y| = 1$ or 3^4. If 1, $X \cong 3 \times Sp(4,3) \times 2$ and we have a contradiction to $\vartheta_3 \sim \vartheta_5$ in C since ϑ_3 is in the derived group of $C_C(\vartheta_3) \cong SU(3,2) \times 2^{1+2} : 3$. We conclude that $|Y| = 3^4$. Since z inverts Y and Y is an irreducible module for X, either Y is extraspecial or elementary abelian. If elementary abelian, it is the direct sum, as X-modules, of the $+1$ and -1 eigenspaces for the action of z. But then, ϑ_5 is not in the commutator subgroup of its centralizer, a contradiction as above. We conclude that $O(X) \cong 3^{1+4}$, extraspecial; it has plus type, by irreducibility of the action of X.

The centralizer, $C(\vartheta_2) \sim C(\vartheta_4)$ is a little troublesome to pin down. The group modulo the center turns out to be not simple ($\cong PSU(4,3)$:2). We know the centralizer of an involution in it is $\cong 2 \times 3 \times Sp(4,3)$; the group $X := C(\vartheta_4)/\langle -1, O(C(\vartheta_4))\rangle$ then has an involution (the image of z) with centralizer isomorphic to $Sp(4,3)$; a classification result (Appendix 10B) identifies X. A more "elementary" argument here would be to analyze the sublattices invariant under $C(\vartheta_4)$ and identify them with the $\mathbb{Z}\left[e^{2\pi 1/3}\right]$-lattices discussed in [Lin71a, b].

For $C(\vartheta_1)$, we have $C_C(\vartheta_1) \cong \left[2_+^{1+6} \times 2\right].\Omega^-(6,2)$ and we quote the characterization of the Suzuki group by centralizer of involution in [Pat]; see Appendix 10B. All we need to do is show that z is not weakly closed in this group; this is so since all involutions of $O_2(C) \backslash Z(O_2(C))$ are conjugate to $\varepsilon_{\mathcal{O}}$ or $\varepsilon_{\mathcal{O}+\Omega}$. As for the previous centralizer, identification of Λ via the $\mathbb{Z}\left[e^{2\pi i/3}\right]$-lattices of [Lin71b] would give another possibility for identifying this group. □

(10.36) Remark. (i) The group $\mathcal{TG}$: $Aut(\mathcal{TG})$ is the analogue for the prime 3 of N_{24}. Unlike the situation for involutions, where every involution has a conjugate in $O_2(N_{24})$, $O_3(\mathcal{TG}: Aut(\mathcal{TG}))$ has representatives of only three of the four classes. The one missing is the class with spectrum $1^8\omega^8\bar{\omega}^8$; this one *must* be missing, since the ternary construction makes it clear that elements of $O_3(\mathcal{TG}: Aut(\mathcal{TG}))$ have eigenvalue multiplicities divisible by 6.

(ii) The group $C/O_2(C) \cong D_4(2)$ has three eight dimensional irreducibles, and these form an orbit under the triality outer automorphism. In $\Lambda/2\Lambda$, it can be shown that all three occur, as compostion factors obtained from the lower central series with respect to the action of $O_2(C)$ on $\Lambda/2\Lambda$.

(10.36) Proposition. *In* .223, *there is one class of involutions and the centralizer of an involution has the form* $2\cdot Alt_8$.

Proof. We let $w := \varepsilon_{\mathcal{O}}$. Then w is in the stabilizer of any triangle supported over $S := \mathcal{O}+\Omega$. If we take an octad $\mathcal{O}'$ disjoint from $\mathcal{O}$, define $\mathcal{O}'' := S+\mathcal{O}'$ and take $i \in \mathcal{O}'$ and $j \in \mathcal{O}''$, we get the triangle $\Delta := (x,y,z)$, where $y = -\alpha_{ij}$, $z = 1/2\alpha_{\mathcal{O}'}$ and $x = -y-z$. Observe that $Stab_{O_2(C(w))}(\Delta) = \langle w\rangle$.

Let $H := O_2(N_{24})G(T)$, where T is the above trio of octads; $G(T) = R(T)L(T)$ (6.10). Clearly, $H \cap O_2(N_{24}) = \langle \varepsilon_A \mid A \in \mathcal{G},\ A$ avoids $\mathcal{O}'$ and $j\rangle \cong 2^4$. Let L_1 be the subgroup of $L(T) \cong GL(3,2) \times \Sigma_3$ fixing each octad in T. If $g \in L_1, g$ stabilizes z but may move y; in any case, $y^g = y^r$, for some $r \in R(T)$. Since $z = z^r$, we get gr in the triangle stabilizer. Thus, $C(w) \cap Stab(\Delta)$ contains a subgroup K of the form $2.2^3.GL(3,2)$; also, $O_2(K)$ is an indecomposable module for $K/O_2(K)$, as can be seen by recalling the action of the octad stabilizer on $\langle \varepsilon_A \mid A \in \mathcal{G}, A \cap \mathcal{O} = \emptyset\rangle$.

We get another subgroup in $C(w)$ as follows. Take the triangle $\Delta' := (x,y,z)$, where x and y are supported at octads $\mathcal{O}^*$ and $\mathcal{O}^{**}$ which avoid $\mathcal{O}$ and meet in a 2-set where their coordinates have opposite signs. This gives us a subgroup isomorphic to Σ_6 in $C(w) \cap Stab(\Delta')$. We now observe that the M_{22}-subgroup of .223 has one conjugacy class of involutions (6.42.4); therefore, if Γ is any triangle fixed by $w, C(w) \cap Stab(\Gamma)$ has order divisible by the least common multiple of $|2.2^3.GL(3,2)|$ and $6!$, i.e., by $2^7 3^2 5.7$, which is the order of $|2\cdot Alt_8|$.

We now show that $2^7 3^2 5.7$ is the right order. Let L be ths sublattice of points fixed by w; $L \cong \sqrt{2}L_{E_8}$. Then, in $L/2L$, an 8-dimensional space with a nonsingular quadratic form of maximal Witt index; the image of Γ spans a nonsingular subspace of maximal index and so its stabilzer in the orthogonal group is isomorphic to $O^+(6,2) \cong \Sigma_8$. Since $Stab_{O_2(C(w))}(\Delta) = \langle w \rangle, (C(w) \cap Stab(\Delta))/\langle w \rangle$ is isomorphic to a subgroup of $O^+(6,2)$, which, by order considerations, is $\Omega^+(6,2) \cong Alt_8$. Thus, $C(w) \cap Stab(\Delta)$ is a central extension $2.Alt_8$. By the above observation that $O_2(K)$ is an indecomposable module for $K/O_2(K)$, we conclude that the extension is nonsplit, whence is the covering group of Alt_8 (2.14). □

(10.37) Exercise. Identify $C(w) \cap Stab(\Delta)$ by suitable generators and relations (2.27.3).

(10.38) Exercise. Devise a program to classify the elements of order 5 in G and to obtain their centralizers. Here are some suggestions.

(10.38.1) Show that if $g \in M_{24}$ has order 5, it has cycle shape $1^4 5^4$ and centralizer $2^4 : Alt_4$ in N_{24}. It may be assumed to lie in $C = C(z)$.

(10.38.2) Show that every element of order 5 in G centralizes an involution. Do so by studying the 5-local $2 \times 5^{1+2} : [4 \circ SL(2,5)].4$. Notice that the Sylow 5-subgroup has exponent 5.

(10.38.3) Show that C has three classes of elements order 5 (imitate the treatment of elements of order 3) and that their G-centralizers have the shapes $5 \times SL(2,5) \circ SL(2,5) : 2$, $5^{1+2} : [4 \circ SL(2,5)] \times 2$ and $5 \times 4\cdot HJ$, where HJ is the simple group of Hall-Janko of order $604800 = 2^7 3^3 5^2 7$ and $4\cdot HJ$ is the central product of a cyclic group of order 4 with the double cover of HJ (2.14). For the latter, one needs to see that the C-centralizer has the form $2_-^{1+4} : Alt_5$; see Appendix 10B.

(10.38.4) Give a pentacode (3.22) construction of Λ so that we get an embedding of a group $5^3 : [4 \times Alt_5].2\times 2$. Work out the G-conjugacy classes of elements of order 5 in this group.

(10.39) Exercise. Get the classes of elements of order 7 in G. Show that there are two, with centralizers of the form $7 \times 2 \cdot Alt_7$ and $7 \times 2 \times \Sigma_4$. Get the heptacode (3.22) into the analysis by building a 7-version of Λ.

(10.40) Exercise. Prove that the pointwise stabilizer of a triangle of type 333 is of the form $3^5.M_{11}$ and that the global stabilizer has the form $3^5 : M_{11} : \Sigma_3 \cong 3^6 : [2 \times M_{11}] \times 2$.

(10.41) Exercise. (i) Prove that Co_3 has two classes of involutions, with centralizers of shapes $2 \cdot Sp(6,2)$ and $2 \times M_{12}$.

(ii) Prove that Co_2 has an involution with centralizer of the form $2_+^{1+8} . Sp(6,2)$ and that the action of $Sp(6,2)$ on the Frattini factor is irreducible (and interpret this as a suitable spin module; get triality for $D_4(2)$ into the analysis). Prove also that this extension splits. (This exercise goes back to a nice observation of Rudvalis: Co_2 contains $.222 \cong PSU(6,2)$ extended by a field automorphism, whose fixed point subgroup is isomorphic to $Sp(6,2)$). Show that there are three classes of involutions in this group and get their centralizers.

(10.42) Exercise. Show that Co_3 contains a 2-local subgroup $2^4 \cdot GL(4,2)$, and prove nonsplitting. It is worth noticing that the subgroup $2^4 : Alt_7$ splits; in fact, O'Nan observed that all cohomology for Alt_7 on this module is zero; the result of [Curr] [Rob] may be applied here to a 3-local subgroup of Alt_7 which has odd index.

(10.43) Exercise. There is an increasing sequence of subgroups $H_n \cong 2 \cdot Alt_n$, for $n = 2, \dots, 9$ such that $C(H_n)$ is isomorphic to, respectively, $Co_0, 6 \cdot \mathrm{Suz}, 2 \cdot G_2(4), 2 \cdot HJ, 2 \times PSU(3,3), 2 \times PSL(3,2), 2 \times Alt_4, \mathbb{Z}_6$. Verify some of these statements with the results and methods of Section 10. See also treatments of some of these groups by [Tits80a, b].

(10.44) Exercise. (i) Show that .223 has an outer automorphism of order 2 with fixed point subgroup isomorphic to M_{11}. (ii) Show that .233 has an outer automorphism with fixed point subgroup isomorphic to Σ_8.

(10.45) A simple group of order 604801. There was an amusing story in connection with the basic reference [HW] on the Hall-Janko simple group. The title of [HW] was "A simple group of order 604800". The next integer, 604801, happens to be a prime. Shortly after [HW] was published, a very short article called "A simple group or order 604801" was sent to Marshall Hall, Jr., who was an editor of The Journal of Algebra. David Wales confirms this story.

Appendices 10A–F

Appendix 10A. D. G. Higman's Theory of Rank 3 Permutation Groups

We mention a few highlights of the theory of rank 3 permutation groups, as given in [Hig64, 71].

(10A.1) Definition. A permutation group has rank r if it is transitive and a point stabilizer has r orbits.

The rank of a transitive group on a set X is the same as the number of orbits on $X \times X$ with diagonal action. A transitive permuation group is doubly transitive iff its rank is 2. Thus, rank 3 is the next step beyond the multiply transitive situation.

(10A.2) Definition. Suppose that G is a rank 3 group on a set X. Let R be an orbit on $X \times X$ which is not the diagonal. Assume that R is symmetric ($(x, y) \in R$ iff $(y, x) \in R$), which is the case if $|G|$ is even. The *associated graph on* X is R.

(10A.3) Notation. For $a \in X$, let $\Delta(a) := \{b \in X \mid (a, b) \in R\}$ and let $\Gamma(a) := X \setminus (\{a\} \cup \Delta(a))$.

Notice that, for $a \in X$, $g \in G$, $\Delta(a)^g = \Delta(a^g)$ and $\Gamma(a)^g = \Gamma(a^g)$.

(10A.4) Notation. $n := |X|$, $k := |\Delta(a)|$, $l := |\Gamma(a)|$,

$$\lambda := |\Delta(a) \cap \Delta(b)|, \quad \text{for } b \in \Delta(a) ,$$

$$\mu := |\Delta(a) \cap \Delta(b)|, \quad \text{for } b \in \Gamma(a) .$$

Thus, $n = k + l + 1$.

If we replace R by the other nontrivial orbit on $X \times X$ (equivalently, if we interchange Δ and Γ in the above), the analogous parameters λ_1 and μ_1 which arise satisfy $\lambda_1 = l - k + \mu - 1$ and $\mu_1 = l - k + \lambda - 1$.

(10A.5) Lemma. $\mu l = k(k - \lambda + 1)$.

(10A.6) Notation. Suppose that the permutation character for G on X decomposes into the principal character plus nonprincipal characters of degrees f_2 and f_3 [Hig64].

(10A.7) Lemma. *Suppose that $|G|$ is even. Then either*

(i) $k = l$, $\mu = \lambda + 1 = k/2$ *and* $f_2 = f_3 = k$; *or*
(ii) $d := (\lambda - \mu)^2 + 4(k - \mu)$ *is a square and*
(a) *if n is even,* $\sqrt{d}$ *divides* $2k + (\lambda - \mu)(k + l)$ *while* $2\sqrt{d}$ *does not, while*
(b) *if n is odd,* $2\sqrt{d}$ *divides* $2k + (\lambda - \mu)(k + l)$.

In case (b), $f_2 \neq f_3$ and the eigenvalues of the incidence matrix for R are integers.

(10A.8) (Parameters for some rank 3 groups). We thank Donald G. Higman for supplying the information in Table 10A.1 about particular rank 3 groups. Here, $\{f, g\} = \{f_1, f_2\}$ and r, s are eigenvalues of an adjacency matrix.

(10A.9) Example. Here is a sequence of permutation groups: Σ_4, $PSL(2, 7)$, $U(3, 3)$, HJ, $G_2(4)$, Suz, of respective degrees 4, 7, 28, 100, 416, 1782. The first group has rank 2 and the rest are rank 3. Each group is a point stabilizer in the next. There is no continuation of this sequence to a rank 3 group with Suz as a point stabilizer. [Suz68, 72]

Table 10A1. Examples of rank 3 simple groups and numerical invariants

Group	Stabilizer	n	k	l	λ	μ	r	s	f	g	[Atlas]
HJ	$U_3(3)$	100	36	63	14	12	6	−4	36	63	[42]
HS	M_{22}	100	22	77	0	6	2	−8	22	77	[82]
McL	$U_4(3)$	275	112	162	30	56	2	−28	252	22	[100]
Suz	$G_2(4)$	1782	416	1365	100	96	20	−16	780	1001	[131]
Co_2	$U_6(2){:}2$	2300	891	1408	378	324	63	−9	275	2024	[154]
Ru	${}^2F_4(2)$	4060	2304	1255	1328	1208	64	−16	783	3276	[126]
Fi_{22}	$2.U_6(2)$	3510	693	2816	180	126	63	−9	429	3080	[163]
Fi_{23}	$2.Fi_{22}$	31671	3510	28160	693	351	351	−9	782	30888	[177]
Fi_{24}	Fi_{23}	306936	31671	275264	3510	3240	351	−81	57477	249485	[207]
Fi_{23}	$O_7(3)$	14080	10920	3159	8408	8680	8	−280	13650	429	[163]
Fi_{23}	$O_8^+(3){:}S_3$	137632	109200	28431	86600	86800	80	−280	106743	30888	[177]

Appendix 10B. Classification Theorems

Below is a summary of the classification theorems used in this chapter. The general form of such a result is as follows: if X is a finite group with involution t such that $C_X(t)$ is as in column 2, then the nontrivial possibilities for X are as listed. The trivial cases are excluded, i.e., $X = O(X)C_X(t)$, with t inverting $O(X)$. In some cases, $X = YY^t\langle t\rangle$ is possible (the "wreathed case"); if so, t must be a direct factor of $C_X(t)$.

Table 10B1. Characterizations of a few simple groups by centralizer of involution

Simple group	Centralizer	Reference
Alt_8, Alt_9	$2^3{:}Sym_4$	[Held]
HJ	$2_-^{1+4}{:}Alt_5$	[HW]
HS	$4\circ 2^{1+4}{:}Sym_5$	[JW69]
McL	$2{}^{\cdot}Alt_8$	[JW71]
Suz	$2_-^{1+6}{}^{\cdot}\Omega^-(6,2)$	[PW2]
Co_2	$2_+^{1+8}{}^{\cdot}Sp(6,2)$	[SmiF]
Co_3	$2{}^{\cdot}Sp(6,2)$	[Fen]
Co_1	$2_+^{1+8}{}^{\cdot}\Omega^+(8,2)$	[Pat]

Note: the chief factors of the centralizers for Suz and Co_1 do not determine their respective isomorphism types; part of the characterization work is to decide which isomorphism type is allowed as the centralizer of involution in a simple group.

Appendix 10C. Tables of Stabilizers

The following table appeared in [Conway1968].

Table 10C1. The vectors of $\Lambda_2, \Lambda_3, \Lambda_4$

Class	Λ_2^2	Λ_2^3	Λ_2^4	Λ_3^2	Λ_3^3	Λ_3^4	Λ_3^5
Shape	$(2^8 0^1 6)$	$(3\ 1^2 3)$	$(4^2 0^{22})$	$(2^{12} 0^{12})$	$(3^3 1^{21})$	$(4\ 2^8 0^{15})$	$(5\ 1^{23})$
No.	$2^7.759$	$2^1 2.24$	$2^2\binom{24}{2}$	$2^{11}.2576$	$2^{12}\binom{24}{3}$	$2^6.759.16$	$2^{12}.24$

Class	Λ_4^{2+}	Λ_4^{2-}	Λ_4^3	Λ_4^4	Λ_4^{4+}	Λ_4^{4-}	Λ_4^5	Λ_4^6	Λ_4^8
Shape	$(2^{16} 0^8)$	$(2^{16} 0^8)$	$(3^5 1^{19})$	$(4^4 0^{20})$	$(4^2 2^8 0^{14})$	$(4\ 2^{12} 0^{11})$	$(5\ 3^2 1^{21})$	$(6\ 2^7 0^{16})$	$(8\ 0^{23})$
No.	$2^1 1.759$	$2^1 1.759.15$	$2^{12}\binom{24}{5}$	$2^4\binom{24}{4}$	$2^9.759.\binom{16}{2}$	$2^{12}.2576.12$	$2^1 2\binom{24}{3}.3$	$2^7.759.8$	$2^1.24$

Table 10C2

Name	Order	Structure	Name	Order	Structure
$\cdot 0$	$2^{22}\,3^9\,5^4\,7^2\,11.13.23$	New perfect	$\cdot 222$	$2^{15}\,3^6\,5.7.11$	$PSU_6(2)$
$\cdot 1$	$2^{21}\,3^9\,5^4\,7^2\,11.13.23$	New simple	$\cdot 322$	$2^7\,3^6\,5^3\,7.11$	M^c
$\cdot 2$	$2^{18}\,3^6\,5^3\,7.11.23$	New simple	$\cdot 332$	$2^9\,3^2\,5^3\,7.11$	HS
$\cdot 3$	$2^1 0\,3^7\,5^3\,7.11.23$	New simple	$\cdot 333$	$2^4\,3^7\,5.11$	$3^5.M_{11}$
$\cdot 4$	$2^{18}\,3^2\,5.7.11.23$	$2^{11}\,M_{23}$	$\cdot 422$	$2^{17}\,3^2\,5.7.11$	$2^{10}.M_{22}$
$\cdot 5$	$2^8\,3^6\,5^3\,7.11$	$M^c.2$	$\cdot 432$	$2^7\,3^2\,5.7.11.23$	M_{23}
$\cdot 6_{22}$	$2^{16}\,3^6\,5.7.11$	$PSU_6(2).2$	$\cdot 433$	$2^{10}\,3^2\,5.7$	$2^4.A_8$
$\cdot 6_{32}$	$2^{10}\,3^3\,5.7.11.23$	M_{24}	$\cdot 442$	$2^{12}\,3^2\,5.7$	$2^{1+8}A_7$
$\cdot 7$	$2^9\,3^2\,5^3\,7.11$	HS	$\cdot 443$	$2^7\,3^2\,5.7$	$M_{21}.2$
$\cdot 8_{22}$	$2^{18}\,3^6\,5^3\,7.11.23$	$\cdot 2$	$\cdot 522$	$2^7\,3^6\,5^3\,7.11$	M^c
$\cdot 8_{32}$	$2^7\,3^6\,5^3\,7.11$	M^c	$\cdot 532$	$2^8\,3^6\,5.7$	$PSU_4(3).2$
$\cdot 8_{42}$	$2^{15}\,2^2\,5.7$	$2^5.2^4.A_8$	$\cdot 533$	$2^4\,3^2\,5^3\,7$	$PSU_3(5)$
$\cdot 9_{33}$	$2^5\,3^7\,5.11$	$3^6.M_{11}.2$	$\cdot 542$	$2^7\,3^2\,5.7.11$	M_{22}
$\cdot 9_{42}$	$2^7\,3^2\,5.7.11.23$	M_{23}	$\cdot 633$	$2^6\,3^3\,5.11$	M_{12}
$\cdot 10_{33}$	$2^{10}\,3^2\,5^3\,7.11$	$HS.2$	$*2 = !2$	$2^{19}\,3^6\,5^3\,7.11.23$	$(\cdot 2)\times 2$
$\cdot 10_{42}$	$2^{17}\,3^2\,5.7.11$	$2^{10}.M_{22}$	$*3 = !3$	$2^{11}\,3^7\,5^3\,7.11.23$	$(\cdot 3)\times 2$
$\cdot 11_{43}$	$2^{10}\,3^2\,5.7$	$2^4.A_8$	$*4$	$2^{19}\,3^2\,5.7.11.23$	$(\cdot 4)\times 2$
$\cdot 11_{52}$	$2^8\,3^6\,5.7$	$PSU_4(3).2$	$!4$	$2^{22}\,3^3\,5.7.11.23$	$2^{12}.M_{24}$
			$!333$	$2^7\,3^9\,5.11$	$3^6.2.M_{12}$
			$!442$	$2^{15}\,3^4\,5.7$	$2^{1+8}A_9$

HS, M^c, p^n denote respectively the Higman–Sims group, the McLaughlin group, and the elementary group of order p^n. $A.B$ denotes an extension of the group A by the group B. The notation is otherwise standard.

Appendix 10D. Maximal Subgroups

Maximal subgroups of M_{24} are treated in Chapter 6. For other simple groups, we refer to [Atlas], which summarizes literature on this subject.

We want to mention two updates about maximal subgroups. One is that the Monster has a 7-local maximal subgroup not recorded in the Atlas; it is described in [Ho]. The Atlas lists the large simple Conway group as having 24 conjugacy classes of maximal subgroups (p. 183). Since then, some errors were noted and the number has shrunk to 22. Robert A. Wilson graciously supplied this recent update (February, 1997):

"The corrections are printed in Appendix 2 of the Atlas of Brauer characters ([AtlasBC] p. 304) – two of the 3-locals turn out to be wrong by various factors of 2, and either do not exist or are not maximal – details can be found in my paper in J. Alg. 113 (1988) 261–262 [Wil88]. There was also a similar mistake in the 5-locals which means that the group described as $5^2{:}4A_5$ actually only has half that order, and shape $5^2{:}2A_5$ – this has not been published. In particular, it seems that there are just 22 classes of maximal subgroups of Co_1 – unless perhaps there are some more mistakes waiting to be found.

Rob Wilson."

Appendix 10E. Nick Patterson's Thesis

Due to its contribution to our knowledge of the Conway groups, we give a summary of the doctoral thesis of Nick Patterson:

On Conway's Group .0 and some Subgroups, by Nicholas James Patterson of Trinity Hall in the University of Cambridge, 1974.

Chapter 1. The conjugacy classes of $N = N_{24}$ (originally worked out by J.H. Conway and M.J.T. Guy).

The elements of N of N (a monomial group) are assigned to "signed permutations". Fusion from N to $G = Co_0$ is studied, starting with elements of orders 2, 4, 8; the results are summarized in Appendix C. Then elements of orders 13, 7, 5 and 3. Next, elements of orders 9, 6. The remaining cases are done in Appendix C.

Chapter 2. The main theorem is the following.

Theorem. *Let G be a group such that $Z^*(G) = 1$. Suppose there is an involution $t \in G$ such that*

(i) $C_G(t) \cong 2^{1+6}PSp(4,3)$ *(recall that* $PSp(4,3) \cong \Omega^-(6,2) \cong PSU(4,2)$*);*
(ii) $C_G(O_2(C_G(t))) = \langle t \rangle$.

Then, G is isomorphic to Suzuki's sporadic group.

The hypotheses (i) and (ii) imply immediately that $C_G(t)$ have one of two isomorphism types; in the language of (2.21), these are a standard and a twisted partial holomorph.

First, it is shown that $|G|$ is determined. Much local information is also obtained.

Secondly, it is shown that if Q is the conjugacy class of elements of order 3 in G whose elements have centralizer of the form $3 \cdot PSU(4,3)$, then two noncommuting elements of order 3 in Q generate a subgroup isomorphic to $SL(2,3)$, $SL(2,5)$ or $PSL(2,5)$. A result of Bernd Stellmacher then identifies G as the Suzuki sporadic group.

Thirdly, it is shown that if $C_G(t)$ is a standard holomorph, then $G = C_G(t)$; hence, Suzuki's group involves the twisted holomorph as an involution centralizer.

Chapter 3.

Theorem. *Let G be a group such that* $Z^*(G) = 1$. *Suppose there is an involution* $t \in G$ *such that* (i) $C_G(t) \cong 2^{1+8}\Omega^+(8,2)$; (ii) $C_G(O_2(C_G(t))) = \langle t \rangle$. *Then, G is isomorphic to* .1.

The style of proof resembles that of the characterization of Suzuki's group. Local analysis is a little harder due to larger Sylow groups. For instance, if S is a Sylow 2-group of $\Omega^+(8,2)$, then $SCN_7(S) = \emptyset$ and $SCN_6(S) \neq \emptyset$ (Definition: $SCN_k(P)$ is the set of self centralizing normal abelian subgroups of rank k in the p-group P). If T is a Sylow 2-group of $C_G(t)$, then $SCN_{12}(T) = \emptyset$ and $SCN_{11}(T) \neq \emptyset$. (For more details on 2-structure of the Conway groups, the reader is referred to [Curtis]). Eventually, a subgroup $N \cong N_{24}$ is constructed and information gotten about its involutions. The order of G is obtained by the Thompson group order formula [Held]. Using a good class of elements of order 3 in G (those with centralizer of the form $3 \cdot Suz$), G is identified by use of Stellmacher's theorem. Finally, G has the twisted holomorph for $C_G(t)$ and as in Chapter 2, it is shown that if $C_G(t)$ is standard, then $G = C_G(t)$.

Chapter 4. A theorem of Rudvalis is proved here, that $H := .2$ contains a class of involutions K such that if x, y are any two elements of K, then $|xy|$ is 1,2,3 or 4, and all of these numbers occur. So, K is a set of so-called $\{3,4\}$-transpositions. A corollary is Conway's identification of .222 as $PSU(6,2)$, which is a group generated by a class of 3-transpositions [Fi][Fi.I].

Appendix A sets up notation for conjugacy classes in N. Appendices B, C and D contain centralizer and power map information for N, .0 and .1.

Appendix 10F. Character Tables

Table 10F1. $PSL(2,7) = PSL(3,2)$; Order = 168 = $2^3.3.7$, Mult = 2, Out = 2

	;	@	@	@	@	@	@	;	;	@	@	@	@
		168	8	3	4	7	7			6	3	4	4
	p power		A	A	A	A	A			A	AB	A	A
	p′ part		A	A	A	A	A			A	AB	A	A
	ind	1A	2A	3A	4A	7A	B**	fus	ind	2B	6A	8A	B*
χ_1	+	1	1	1	1	1	1	:	++	1	1	1	1
χ_2	o	3	−1	0	1	b7	**	│	+	0	0	0	0
χ_3	o	3	−1	0	1	**	b7	│					
χ_4	+	6	2	0	0	−1	−1	:	++	0	0	r2	−r2
χ_5	+	7	−1	1	−1	0	0	:	++	1	1	−1	−1
χ_6	+	8	0	−1	0	1	1	:	++	2	−1	0	0
	ind	1	4	3	8	7	7	fus	ind	4	12	16	16
		2		6	8	14	14				12	16	16
χ_7	o	4	0	1	0	−b7	**	│	−	0	0	0	0
χ_8	o	4	0	1	0	**	−b7	│					
χ_9	−	6	0	0	r2	−1	−1	:	−−	0	0	y16	*3
χ_{10}	−	6	0	0	−r2	−1	−1	:	−−	0	0	*5	y16
χ_{11}	−	8	0	−1	0	1	1	:	−−	0	r3	0	0

Table 10F2. Mathieu group M_{11}: Order = 7920 = $2^4 3^2 5.11$, Multiplier = 1, Out = 1

	;	@	@	@	@	@	@	@	@	@	@
		7920	48	18	8	5	6	8	8	11	11
	p power		A	A	A	A	AA	A	A	A	A
	p′ part		A	A	A	A	AA	A	A	A	A
	ind	1A	2A	3A	4A	5A	6A	8A	B**	11A	B**
χ_1	+	1	1	1	1	1	1	1	1	1	1
χ_2	+	10	2	1	2	0	−1	0	0	−1	−1
χ_3	o	10	−2	1	0	0	1	i2	−i2	−1	−1
χ_4	o	10	−2	1	0	0	1	−i2	i2	−1	−1
χ_5	+	11	3	2	−1	1	0	−1	−1	0	0
χ_6	o	16	0	−2	0	1	0	0	0	b11	**
χ_7	o	16	0	−2	0	1	0	0	0	**	b11
χ_8	+	44	4	−1	0	−1	1	0	0	0	0
χ_9	+	45	−3	0	1	0	0	−1	−1	1	1
χ_{10}	+	55	−1	1	−1	0	−1	1	1	0	0

Table 10F3. Mathieu group M_{12}: Order = 95,040 = $2^6 3^3 5.11$, Multiplier = 2, Out = 2

	;	@	@	@	@	@	@	@	@	@	@	@	@	@	@	@	;	;	@	@	@	@	@	@	@	@	@
		95040	240	192	54	36	32	32	10	12	6	8	8	10	11	11			120	24	12	6	10	10	6	6	6
p power			A	A	A	A	B	B	A	BA	AB	A	B	AA	A	A			A	B	A	BC	AC	AC	AD	BC	BC
p' part			A	A	A	A	A	A	A	BA	AB	A	A	AA	A	A			A	A	A	BC	AC	AC	BD	AC	AC
ind		1A	2A	2B	3A	3B	4A	4B	5A	6A	6B	8A	8B	10A	11A	B**	fus	ind	2C	4C	4D	6C	10B	C*	12A	12B	C*
χ_{1}	+	1	1	1	1	1	1	1	1	1	1	1	1	1	1	1	:	++	1	1	1	1	1	1	1	1	1
χ_{2}	+	11	−1	3	2	−1	−1	3	1	−1	0	−1	1	−1	0	0	|	+	0	0	0	0	0	0	0	0	0
χ_{3}	+	11	−1	3	2	−1	3	−1	1	−1	0	1	−1	−1	0	0	|										
χ_{4}	∘	16	4	0	−2	1	0	0	1	1	0	0	0	−1	b11	**	|	+	0	0	0	0	0	0	0	0	0
χ_{5}	∘	16	4	0	−2	1	0	0	1	1	0	0	0	−1	**	b11	|										
χ_{6}	+	45	5	−3	0	3	1	1	0	−1	0	−1	−1	0	1	1	:	++	5	−3	1	−1	0	0	1	0	0
χ_{7}	+	54	6	6	0	0	2	2	−1	0	0	0	0	1	−1	−1	:	++	0	0	0	0	r5	−r5	0	0	0
χ_{8}	+	55	−5	7	1	1	−1	−1	0	1	1	−1	−1	0	0	0	:	++	5	1	−1	−1	0	0	−1	1	1
χ_{9}	+	55	−5	−1	1	1	3	−1	0	1	−1	−1	1	0	0	0	|	+	0	0	0	0	0	0	0	0	0
χ_{10}	+	55	−5	−1	1	1	−1	3	0	1	−1	1	−1	0	0	0	|										
χ_{11}	+	66	6	2	3	0	−2	−2	1	0	−1	0	0	1	0	0	:	++	6	2	0	0	1	1	0	−1	−1
χ_{12}	+	99	−1	3	0	3	−1	−1	−1	−1	0	1	1	−1	0	0	:	++	1	−3	−1	1	1	1	−1	0	0
χ_{13}	+	120	0	−8	3	0	0	0	0	0	1	0	0	0	−1	−1	:	++	0	0	0	0	0	0	0	r3	−r3
χ_{14}	+	144	4	0	0	−3	0	0	−1	1	0	0	0	−1	1	1	:	++	4	0	2	1	−1	−1	−1	0	0
χ_{15}	+	176	−4	0	−4	−1	0	0	1	−1	0	0	0	1	0	0	:	++	4	0	−2	1	−1	−1	1	0	0
ind		1	4	2	3	3	4	4	5	12	6	8	8	20	11	11	fus	ind	2	4	8	6	10	10	24	12	12
		2		2	6	6			10		6	8	8	20	22	22				4	8				24	12	12
χ_{16}	∘	10	0	−2	1	−2	0	0	0	0	1	i2	i2	0	−1	−1	:	oo	0	2	i2	0	0	0	i2	−1	−1
χ_{17}	∘	10	0	−2	1	−2	0	0	0	0	1	−i2	−i2	0	−1	−1	:	oo	0	2	−i2	0	0	0	−i2	−1	−1
χ_{18}	+	12	0	4	3	0	0	0	2	0	1	0	0	0	1	1	:	++	0	0	0	0	0	0	0	r3	−r3
χ_{19}	−	32	0	0	−4	2	0	0	2	0	0	0	0	0	−1	−1	:	oo	0	0	2i2	0	0	0	−i2	0	0
χ_{20}	∘	44	0	4	−1	2	0	0	−1	0	1	0	0	i5	0	0	|	+	0	0	0	0	0	0	0	0	0
χ_{21}	∘	44	0	4	−1	2	0	0	−1	0	1	0	0	−i5	0	0	|										
χ_{22}	∘	110	0	−6	2	2	0	0	0	0	0	i2	−i2	0	0	0	|	+	0	0	0	0	0	0	0	0	0
χ_{23}	∘	110	0	−6	2	2	0	0	0	0	0	−i2	i2	0	0	0	|										
χ_{24}	+	120	0	8	3	0	0	0	0	0	−1	0	0	0	−1	−1	:	++	0	4	0	0	0	0	0	1	1
χ_{25}	∘	160	0	0	−2	−2	0	0	0	0	0	0	0	0	−b11	**	|	+	0	0	0	0	0	0	0	0	0
χ_{26}	∘	160	0	0	−2	−2	0	0	0	0	0	0	0	0	**	−b11	|										

Table 10F4. $PSL(2,23)$: Order = 6,072 = 2^3.3.11.23, Mult = 2, Out = 2

	;	@	@	@	@	@	@	@	@	@	@	@	@	@	@	;	;	@	@	@	@	@	@	@	@	@	@	@	@
		6072	24	12	12	12	11	11	11	11	11	12	12	23	23			22	12	12	11	11	11	11	11	12	12	12	12
	p power		A	A	A	AA	A	A	A	A	A	AA	AA	A	A			A	A	A	CB	DB	EB	AB	BB	BB	AB	BA	AA
	p′ part		A	A	A	AA	A	A	A	A	A	AA	AA	A	A			A	A	A	AB	BB	CB	DB	EB	AA	AA	AB	AB
	ind	1A	2A	3A	4A	6A	11A	B*3	C*2	D*5	E*4	12A	B*	23A	B**	fus	ind	2B	8A	B*	22A	B*3	C*9	D*5	E*7	24A	B*7	C*11	D*5
χ1	+	1	1	1	1	1	1	1	1	1	1	1	1	1	1	:	++	1	1	1	1	1	1	1	1	1	1	1	1
χ2	o	11	−1	−1	1	−1	0	0	0	0	0	1	1	b23	**		+	0	0	0	0	0	0	0	0	0	0	0	0
χ3	o	11	−1	−1	1	−1	0	0	0	0	0	1	1	**	b23														
χ4	+	22	−2	1	−2	1	0	0	0	0	0	1	1	−1	−1	:	++	0	2	2	0	0	0	0	0	−1	−1	−1	−1
χ5	+	22	2	−2	0	2	0	0	0	0	0	0	0	−1	−1	:	++	0	r2	−r2	0	0	0	0	0	r2	r2	−r2	−r2
χ6	+	22	−2	1	2	1	0	0	0	0	0	−1	−1	−1	−1	:	++	0	0	0	0	0	0	0	0	r3	−r3	r3	−r3
χ7	+	22	2	1	0	−1	0	0	0	0	0	r3	−r3	−1	−1	:	++	0	−r2	r2	0	0	0	0	0	y24	*7	*11	*5
χ8	+	22	2	1	0	−1	0	0	0	0	0	−r3	r3	−1	−1	:	++	0	−r2	r2	0	0	0	0	0	*7	y24	*5	*11
χ9	+	23	−1	−1	−1	−1	1	1	1	1	1	−1	−1	0	0	:	++	1	−1	−1	1	1	1	1	1	−1	−1	−1	−1
χ10	+	24	0	0	0	0	y11	*3	*2	*5	*4	0	0	1	1	:	++	2	0	0	y11	*3	*2	*5	*4	0	0	0	0
χ11	+	24	0	0	0	0	*4	y11	*3	*2	*5	0	0	1	1	:	++	2	0	0	*4	y11	*3	*2	*5	0	0	0	0
χ12	+	24	0	0	0	0	*5	*4	y11	*3	*2	0	0	1	1	:	++	2	0	0	*5	*4	y11	*3	*2	0	0	0	0
χ13	+	24	0	0	0	0	*2	*5	*4	y11	*3	0	0	1	1	:	++	2	0	0	*2	*5	*4	y11	*3	0	0	0	0
χ14	+	24	0	0	0	0	*3	*2	*5	*4	y11	0	0	1	1	:	++	2	0	0	*3	*2	*5	*4	y11	0	0	0	0
	ind	1	4	3	8	12	11	11	11	11	11	24	24	23	23	fus	ind	4	16	16	44	44	44	44	44	48	48	48	48
		2		6	8	12	22	22	22	22	22	24	24	46	46				16	16	44	44	44	44	44	48	48	48	48
χ15	o	12	0	0	0	0	1	1	1	1	1	0	0	−b23	**		−	0	0	0	0	0	0	0	0	0	0	0	0
χ16	o	12	0	0	0	0	1	1	1	1	1	0	0	**	−b23														
χ17	−	22	0	−2	r2	0	0	0	0	0	0	r2	r2	−1	−1	:	−−	0	y16	*3	0	0	0	0	0	y16	y16	*3	*3
χ18	−	22	0	−2	−r2	0	0	0	0	0	0	−r2	−r2	−1	−1	:	−−	0	*5	y16	0	0	0	0	0	*5	*5	y16	y16
χ19	−	22	0	1	−r2	−r3	0	0	0	0	0	y24	*7	−1	−1	:	−−	0	*5	y16	0	0	0	0	0	*17	y48	*19	*13
χ20	−	22	0	1	−r2	r3	0	0	0	0	0	*7	y24	−1	−1	:	−−	0	*5	y16	0	0	0	0	0	y48	*17	*13	*19
χ21	−	22	0	1	r2	−r3	0	0	0	0	0	−y24	*7	−1	−1	:	−−	0	−y16	*3	0	0	0	0	0	*5	*11	*17	y48
χ22	−	22	0	1	r2	r3	0	0	0	0	0	*7	−y24	−1	−1	:	−−	0	−y16	*3	0	0	0	0	0	*11	*5	y48	*17
χ23	−	24	0	0	0	0	y11	*3	*2	*5	*4	0	0	1	1	:	−−	0	0	0	*7	y44	*19	*9	*5	0	0	0	0
χ24	−	24	0	0	0	0	*4	y11	*3	*2	*5	0	0	1	1	:	−−	0	0	0	*5	*7	y44	*19	*9	0	0	0	0
χ25	−	24	0	0	0	0	*5	*4	y11	*3	*2	0	0	1	1	:	−−	0	0	0	*9	*5	*7	y44	*19	0	0	0	0
χ26	−	24	0	0	0	0	*2	*5	*4	y11	*3	0	0	1	1	:	−−	0	0	0	*19	*9	*5	*7	y44	0	0	0	0
χ27	−	24	0	0	0	0	*3	*2	*5	*4	y11	0	0	1	1	:	−−	0	0	0	y44	*19	*9	*5	*7	0	0	0	0

Table 10F5. $PSL(3,4) = M_{21}$: Order = 20160 = $2^6 3^2 5 \cdot 7$, Mult = 12×4, Out = $2 \times \Sigma_3$

	;	@	@	@	@	@	@	@	@	@	@	;	;	@	@	@	@	@	@	;	;	@	@	@	@	@	@	@
		20160	64	9	16	16	16	5	5	7	7			72	8	9	4	4	4			60	21	4	5	5	7	7
	p power		A	A	A	A	A	A	A	A	A			A	A	AB	A	B	C			A	A	BA	BB	AB	BC	AC
	p′ part		A	A	A	A	A	A	A	A	A			A	A	AB	A	A	A			A	A	BA	AB	BB	AC	BC
	ind	1A	2A	3A	4A	4B	4C	5A	B*	7A	B**	fus	ind	2B	4D	6A	8A	8B	8C	fus	ind	3B	3C	6B	15A	B*	21A	B*
χ_{1}	+	1	1	1	1	1	1	1	1	1	1	:	++	1	1	1	1	1	1	:	+oo	1	1	1	1	1	1	1
χ_{2}	+	20	4	2	0	0	0	0	0	−1	−1	:	++	2	−2	2	0	0	0	:	+oo	5	−1	1	0	0	−1	−1
χ_{3}	+	35	3	−1	3	−1	−1	0	0	0	0	:	++	1	1	1	1	−1	−1		+	0	0	0	0	0	0	0
χ_{4}	+	35	3	−1	−1	3	−1	0	0	0	0	:	++	1	1	1	−1	1	−1									
χ_{5}	+	35	3	−1	−1	−1	3	0	0	0	0	:	++	1	1	1	−1	−1	1									
χ_{6}	o	45	−3	0	1	1	1	0	0	b7	**		+	0	0	0	0	0	0	:	ooo	0	3	0	0	0	b7	**
χ_{7}	o	45	−3	0	1	1	1	0	0	**	b7									:	ooo	0	3	0	0	0	**	b7
χ_{8}	+	63	−1	0	−1	−1	−1	−b5	*	0	0		+	0	0	0	0	0	0	:	+oo	3	0	−1	−b5	*	0	0
χ_{9}	+	63	−1	0	−1	−1	−1	*	−b5	0	0									:	+oo	3	0	−1	*	−b5	0	0
χ_{10}	+	64	0	1	0	0	0	−1	−1	1	1	:	++	8	0	−1	0	0	0	:	+oo	4	1	0	−1	−1	1	1
	ind	1	2	3	4	4	4	5	5	7	7	fus	ind	2	4	6	8	8	8									
		2	2	6	4			10	10	14	14					6	8	8	8									
χ_{11}	o	10	2	1	2	0	0	0	0	b7	**		+	0	0	0	0	0	0									
χ_{12}	o	10	2	1	2	0	0	0	0	**	b7																	
χ_{13}	+	28	−4	1	0	0	0	−b5	*	0	0		+	0	0	0	0	0	0									
χ_{14}	+	28	−4	1	0	0	0	*	−b5	0	0																	
χ_{15}	+	36	4	0	0	0	0	1	1	1	1	:	++	0	0	0	2	0	0									
χ_{16}	+	64	0	1	0	0	0	−1	−1	1	1	:	++	0	0	3	0	0	0									
χ_{17}	+	70	−2	−2	2	0	0	0	0	0	0	:	++	0	0	0	0	r2	r2									
χ_{18}	+	90	2	0	−2	0	0	0	0	−1	−1	:	++	0	0	0	0	r2	−r2									
	and	1	2	3	4	4	4	5	5	7	7		and	2	4	6	8	8	8									
	no:	:2	:2	:2		:2		:2	:2	:2	:2		no:			:2	:2	:2	:2									
	and	1	2	3	4	4	4	5	5	7	7		and	2	4	6	8	8	8									
	no:	:2	:2	:2			:2	:2	:2	:2	:2		no:			:2	:2	:2	:2									
	ind	1	2	3	4	8	8	5	5	7	7	fus	ind	2	4	6	8	16	16									
		4	4	12	4			20	20	28	28					6	8	16	16									
		2		6				10	10	14	14																	
		4		12				20	20	28	28																	
χ_{19}	o2	8	0	−1	0	0	0	−b5	*	1	1		o2															
χ_{20}	o2	8	0	−1	0	0	0	*	−b5	1	1																	
χ_{21}	o2	56	0	2	0	0	0	1	1	0	0	*	+															
χ_{22}	o2	64	0	1	0	0	0	−1	−1	1	1	*	+															
χ_{23}	o2	80	0	−1	0	0	0	0	0	b7	**		+2															
χ_{24}	o2	80	0	−1	0	0	0	0	0	**	b7																	
	and	1	2	3	8	4	8	5	5	7	7		and	2	4	6	16	8	16									
	no:	:4	:2	:4		:2		:4	:4	:4	:4		no:			:2	:2	:2	:2									
	and	1	2	3	8	8	4	5	5	7	7		and	2	4	6	16	16	8									
	no:	:4	:2	:4			:2	:4	:4	:4	:4		no:			:2	:2	:2	:2									
	ind	1	2	3	4	8	8	5	5	7	7	fus	ind	2	4	6	8	16	16									
		4	4	12	4			20	20	28	28					6	8	16	16									
		2		6	4			10	10	14	14																	
		4		12	4			20	20	28	28																	
χ_{25}	o2	20	0	2	2	0	0	0	0	−1	−1	*	+															
χ_{26}	o2	28	0	1	−2	0	0	−b5	*	0	0		o2															
χ_{27}	o2	28	0	1	−2	0	0	*	−b5	0	0																	
χ_{28}	o2	36	0	0	2	0	0	1	1	1	1	*	+															
χ_{29}	o2	64	0	1	0	0	0	−1	−1	1	1	*	+															
χ_{30}	o2	80	0	−1	0	0	0	0	0	b7	**		+2															
χ_{31}	o2	80	0	−1	0	0	0	0	0	**	b7																	
	and	1	2	3	8	4	8	5	5	7	7		and	2	4	6	16	8	16									
	no:	:4	:2	:4		:4		:4	:4	:4	:4		no:			:2	:2	:2	:2									
	and	1	2	3	8	8	4	5	5	7	7		and	2	4	6	16	16	8									
	no:	:4	:2	:4			:4	:4	:4	:4	:4		no:			:2	:2	:2	:2									

able 10F6. $PSL(3,4) = M_{21}$: Order = 20160 = $2^6 3^2 5 \cdot 7$, Mult = 12 × 4, Out = 2 × Σ_3

;	;	@	@	@	;	;	@	@	@	@	@	@	;	;	;	;	;	;	@	@	@	@	@	@	;	;	;	;	
		6	3	2			168	8	3	4	7	7							60	3	8	8	5	5					
		BB	CB	BD			A	A	AC	A	AC	BC							A	AD	A	A	BD	AD					
		BB	CB	BD			A	A	AC	A	AC	BC							A	AD	A	A	AD	BD					
us	ind	6C	6D	12A	fus	ind	2C	4E	6E	8D	14A	B**	fus	ind	fus	ind	fus	ind	2D	6F	8E	8F	10A	B*	fus	ind	fus	ind	
:	+oo+oo	1	1	1	:	++	1	1	1	1	1	1	:	++	:	++	:	++	1	1	1	1	1	1	:	++	:	++	χ_1
:	+oo+oo	−1	−1	1	:	++	6	2	0	0	−1	−1	:	++	:	++	:	++	0	0	−2	2	0	0	:	++	:	++	χ_2
:	++	0	0	0	:	++	7	−1	1	−1	0	0	|	+	|	+	:	++	5	−1	1	1	0	0	|	+	|	+	χ_3
|					|	+	0	0	0	0	0	0	:|	++			|	+	0	0	0	0	0	0	:|	++			χ_4
|															:	++											:	++	χ_5
|	+oo	0	0	0	:	oo	3	−1	0	1	b7	**	:	oo	:	oo	|	+	0	0	0	0	0	0	|	+	|	+	χ_6
|					:	oo	3	−1	0	1	**	b7	:	oo	:	oo													χ_7
|	+oo	0	0	0	|	+	0	0	0	0	0	0	|	+	|	+	:	++	3	0	−1	−1	−b5	*	:	++	:	++	χ_8
|																	:	++	3	0	−1	−1	*	−b5	:	++	:	++	χ_9
:	+oo+oo	2	−1	0	:	++	8	0	−1	0	1	1	:	++	:	++	:	++	4	1	0	0	−1	−1	:	++	:	++	χ_{10}
					fus	ind	2	4	6	8	14	14	fus	ind	fus	ind	fus	ind	2	6	8	8	10	10	fus	ind	fus	ind	
							2	4	6	8	14	14							2	6	8	8	10	10					
					:	oo	4	0	1	0	−b7	**					|	+	0	0	0	0	0	0					χ_{11}
					:	oo	4	0	1	0	**	−b7																	χ_{12}
					|	+	0	0	0	0	0	0					:	++	2	−1	0	0	b5	*					χ_{13}
																	:	++	2	−1	0	0	*	b5					χ_{14}
					:	++	6	2	0	0	−1	−1					:	++	6	0	0	0	1	1					χ_{15}
					:	++	8	0	−1	0	1	1					:	++	4	1	0	0	−1	−1					χ_{16}
					:	oo	0	0	0	2i	0	0					:	++	0	0	2r2	0	0	0					χ_{17}
					:	++	6	−2	0	0	−1	−1					:	++	0	0	0	2r2	0	0					χ_{18}
													:	:											:	:			
													:	:											:	:			
															:	:											:	:	
															:	:											:	:	
					fus	ind	2	4	6	8	14	14	fus	ind	fus	ind	fus	ind	2	6	8	8	10	10	fus	ind	fus	ind	
							2	4	6	8	14	14							4	12	8	8	20	20					
																			2	6			10	10					
																			4	12			20	20					
					|	o2											:	oo2	2	−1	0	0	b5	*					χ_{19}
					|												:	oo2	2	−1	0	0	*	b5					χ_{20}
					*	+											:	oo2	6	0	0	0	1	1					χ_{21}
					*	+											:	oo2	4	1	0	0	−1	−1					χ_{22}
					*	o											|	o2	0	0	0	0	0	0					χ_{23}
					*	o																							χ_{24}
													*	*											:	:			
													*	*											:	:			
															*	*											:	:	
															*	*											:	:	
					fus	ind	2	4	6	8	14	14	fus	ind	fus	ind	fus	ind	2	6	8	8	10	10	fus	ind	fus	ind	
							4	4	12	8	28	28							2	6	8	8	10	10					
							2		6	8	14	14																	
							4		12	8	28	28																	
					:	oo2	6	0	0	i+1	−1	−1					*	+											χ_{25}
					|	o2	0	0	0	0	0	0					*	+											χ_{26}
																	*	+											χ_{27}
					:	oo2	6	0	0	−i−1	−1	−1					*	+											χ_{28}
					:	oo2	8	0	−1	0	1	1					*	+											χ_{29}
					:	oo2	4	0	1	0	−b7	**					|	+2											χ_{30}
					:	oo2	4	0	1	0	**	−b7					|												χ_{31}
													:	:											*	*			
													:	:											*	*			
															:	:											*	*	
															:	:											*	*	

Table 10F7. $PSL(3,4)$

		1A	2A	3A	4A	4B	4C	5A	B*	7A	B**			2B	4D	6A	8A	8B	8C			3B	3C	6B	15A	B*	21A	B*
	ind	1	2	3	4	4	4	5	5	7	7	fus	ind	2	4	6	8	8	8	fus	ind	3	9	6	15	15	63	63
		3	6		12	12	12	15	15	21	21			6	12		24	24	24			3		6	15	15	63	63
		3	6		12	12	12	15	15	21	21			6	12		24	24	24			3		6	15	15	63	63
χ_{32}	∘2	15	−1	0	3	−1	−1	0	0	1	1		:∘∘2	3	−1	0	1	−1	−1		∘2	0	0	0	0	0	0	0
χ_{33}	∘2	15	−1	0	−1	3	−1	0	0	1	1		:∘∘2	3	−1	0	−1	1	−1									
χ_{34}	∘2	15	−1	0	−1	−1	3	0	0	1	1		:∘∘2	3	−1	0	−1	−1	−1									
χ_{35}	∘2	21	5	0	1	1	1	1	1	0	0		:∘∘2	3	−1	0	1	1	1		:∘∘∘2	−2i3−3	0	−1	z3	z3	0	0
χ_{36}	∘2	45	−3	0	1	1	1	0	0	b7	**		∘2	0	0	0	0	0	0		:∘∘∘2	0	0	0	0	0	x63	**8
χ_{37}	∘2	45	−3	0	1	1	1	0	0	**	b7										:∘∘∘2	0	0	0	0	0	**8	x63
χ_{38}	∘2	63	−1	0	−1	−1	−1	−b5	*	0	0		∘2	0	0	0	0	0	0		:∘∘∘2	3	0	−1	−b5	*	0	0
χ_{39}	∘2	63	−1	0	−1	−1	−1	*	−b5	0	0										:∘∘∘2	3	0	−1	*	−b5	0	0
χ_{40}	∘2	84	4	0	0	0	0	−1	−1	0	0		:∘∘2	6	2	0	0	0	0		:∘∘∘2	−2i3+3	0	1	z3+1	z3+1	0	0
	ind	1	2	3	4	4	4	5	5	7	7	fus	ind	2	4	6	8	8	8									
		6	6	6	12	12	12	30	30	42	42			6	12	6	24	24	24									
		3	6		12	12	12	15	15	21	21			6	12		24	24	24									
		2	2		4			10	10	14	14						8	8	8									
		3	6		12			15	15	21	21						24	24	24									
		6	6		12			30	30	42	42						24	24	24									
χ_{41}	∘2	6	−2	0	2	0	0	1	1	−1	−1		:∘∘2	0	0	0	0	r2	r2									
χ_{42}	∘2	36	4	0	0	0	0	1	1	1	−1		:∘∘2	0	0	0	2	0	0									
χ_{43}	∘2	42	2	0	2	0	0	b5	*	0	0		∘2	0	0	0	0	0	0									
χ_{44}	∘2	42	2	0	2	0	0	*	b5	0	0																	
χ_{45}	∘2	60	−4	0	0	0	0	0	0	−b7	**		∘2	0	0	0	0	0	0									
χ_{46}	∘2	60	−4	0	0	0	0	0	0	**	−b7																	
χ_{47}	∘2	90	2	0	−2	0	0	0	0	−1	−1		:∘∘2	0	0	0	0	r2	−r2									
	and	1	2	3	4	4	4	5	5	7	7		and	2	4	6	8	8	8									
	no:	:6	:6	:2	:3	:6	:3	:6	:6	:6	:6		no:	:3	:3	:2	:6	:6	:6									
	and	1	2	3	4	4	4	5	5	7	7		and	2	4	6	8	8	8									
	no:	:6	:6	:2	:3	:3	:6	:6	:6	:6	:6		no:	:3	:3	:2	:6	:6	:6									
	ind	1	2	3	4	8	8	5	5	7	7	fus	ind	2	4	6	8	16	16									
		12	12	12	12	24	24	60	60	84	84			6	12	6	24	48	48									
		6	6	6	12	24	24	30	30	42	42			6	12		24	48	48									
		4	4	12	4			20	20	28	28						8	16	16									
		3	6		12			15	15	21	21						24	48	48									
		12	12		12			60	60	84	84						24	48	48									
		2						10	10	14	14																	
		12						60	60	84	84																	
		3						15	15	21	21																	
		4						20	20	28	28																	
		6						30	30	42	42																	
		12						60	60	84	84																	
χ_{48}	∘4	24	0	0	0	0	0	−1	−1	b7	**		∘4															
χ_{49}	∘4	24	0	0	0	0	0	−1	−1	**	b7	*7																
χ_{50}	∘4	48	0	0	0	0	0	−b5	*	−1	−1		∘4															
χ_{51}	∘4	48	0	0	0	0	0	*	−b5	−1	−1	*7																
χ_{52}	∘4	120	0	0	0	0	0	0	0	1	1	*7	∘2															
	and	1	2	3	8	4	8	5	5	7	7		and	2	4	6	16	8	16									
	no:	:12	:6	:4	:3	:6	:3	:12	:12	:12	:12		no:	:3	:3	:2	:6	:6	:6									
	and	1	2	3	8	8	4	5	5	7	7		and	2	4	6	16	16	8									
	no:	:12	:6	:4	:3	:3	:6	:12	:12	:12	:12		no:	:3	:3	:2	:6	:6	:6									
	ind	1	2	3	4	8	8	5	5	7	7	fus	ind	2	4	6	8	16	16									
		12	12	12	12	24	24	60	60	84	84			6	12	6	24	48	48									
		6	6	6	12	24	24	30	30	42	42			6	12		24	48	48									
		4	4	12	4			20	20	28	28						8	16	16									
		3	6		12			15	15	21	21						24	48	48									
		12	12		12			60	60	84	84						24	48	48									
		2			4			10	10	14	14																	
		12			12			60	60	84	84																	
		3			12			15	15	21	21																	
		4			4			20	20	28	28																	
		6			12			30	30	42	42																	
		12			12			60	60	84	84																	
χ_{53}	∘4	36	0	0	2	0	0	1	1	1	1	*7	∘2															
χ_{54}	∘4	48	0	0	0	0	0	−b5	*	−1	−1		∘4															
χ_{55}	∘4	48	0	0	0	0	0	*	−b5	−1	−1	*7																
χ_{56}	∘4	60	0	0	−2	0	0	0	0	−b7	**		∘4															
χ_{57}	∘4	60	0	0	−2	0	0	0	0	**	−b7	*7																
χ_{58}	∘4	84	0	0	2	0	0	−1	−1	0	0	*7	∘2															
	and	1	2	3	8	4	8	5	5	7	7		and	2	4	6	16	8	16									
	no:	:12	:6	:4	:3	:12	:3	:12	:12	:12	:12		no:	:3	:3	:2	:6	:6	:6									
	and	1	2	3	8	8	4	5	5	7	7		and	2	4	6	16	16	8									
	no:	:12	:6	:4	:3	:3	:12	:12	:12	:12	:12		no:	:3	:3	:2	:6	:6	:6									

Table 10F8. $PSL(3,4)$

		6C	6D	12A			2C	4E	6E	8D	14A	B**							2D	6F	8E	8F	10A	B*					
fus	ind	6	18	12	fus	ind	2	4	6	8	14	14	fus	ind	fus	ind	fus	ind	2	6	8	8	10	10	fus	ind	fus	ind	
		6		12																									
		6		12																									
⁝	∘∘2	0	0	0	*	+							↓	∘2	↓	∘2	*	+							↓	∘2	↓	∘2	χ_{32}
⁝					↓	∘2							*	+	*		↓	∘2							*	+	*		χ_{33}
⁝					*								*		*	+	*								*		*	+	χ_{34}
:	∘∘∘∘∘∘∘2	$i3$	0	−1	*	+							*	+	*	+	*	+							*	+	*	+	χ_{35}
⁝	∘∘∘2	0	0	0	*	∘							*	∘	*	∘	↓	+2							↓	+2	↓	+2	χ_{36}
⁝					*	∘							*	∘	*	∘	*								*		*		χ_{37}
⁝	∘∘∘2	0	0	0	↓	∘2							↓	∘2	↓	∘2	*	+							*	+	*	+	χ_{38}
⁝					*								*		*		*	+							*	+	*	+	χ_{39}
:	∘∘∘∘∘∘∘2	$-i3$	0	−1	*	+							*	+	*	+	*	+							*	+	*	+	χ_{40}
					fus	ind	2	4	6	8	14	14	fus	ind	fus	ind	fus	ind	2	6	8	8	10	10	fus	ind	fus	ind	
							2	4	6	8	14	14							2	6	8	8	10	10					
					*	−											*	+											χ_{41}
					*	+											*	+											χ_{42}
					↓	∘2											*	+											χ_{43}
					*												*	+											χ_{44}
					*	∘											↓	+2											χ_{45}
					*	∘											*												χ_{46}
					*	+											*	+											χ_{47}
													*	*											*	*			
													*	*											*	*			
															*	*											*	*	
															*	*											*	*	
					fus	ind	2	4	6	8	14	14	fus	ind	fus	ind	fus	ind	2	6	8	8	10	10	fus	ind	fus	ind	
							2	4	6	8	14	14							4	12	8	8	20	20					
																			2	6			10	10					
																			4	12			20	20					
					**	∘2											↓	∘4											χ_{48}
					**	∘2											*5												χ_{49}
					↓	∘4											*5	∘2											χ_{50}
					**												*5	∘2											χ_{51}
					**	+2											*5	∘2											χ_{52}
													**	**											*5	*5			
													**	**											*5	*5			
															**	**											*5	*5	
															**	**											*5	*5	
					fus	ind	2	4	6	8	14	14	fus	ind	fus	ind	fus	ind	2	6	8	8	10	10	fus	ind	fus	ind	
							4	4	12	8	28	28							2	6	8	8	10	10					
							2		6	8	14	14																	
							4		12	8	28	28																	
					*5	∘2											**	+2											χ_{53}
					↓	∘4											**	+2											χ_{54}
					*5												**	+2											χ_{55}
					*5	∘2											↓	+4											χ_{56}
					*5	∘2											**												χ_{57}
					*5	∘2											**	+2											χ_{58}
													*5	*5											**	**			
													*5	*5											**	**			
															*5	*5											**	**	
															*5	*5											**	**	

Table 10F9. Mathieu group M_{22}: Order = 443,520 = $2^7 3^2 5 \cdot 7 \cdot 11$, Mult = 12, Out = 2

	;	@	@	@	@	@	@	@	@	@	@	@	@	;	;	@	@	@	@	@	@	@	@	@	@
																1									
		443520	384	36	32	16	5	12	7	7	8	11	11			344	320	48	32	6	8	5	6	7	7
	p power		A	A	A	A	A	AA	A	A	A	A	A			A	A	A	A	AB	A	AC	AC	AB	BB
	p′ part		A	A	A	A	A	AA	A	A	A	A	A			A	A	A	A	AB	A	AC	AC	AB	BB
	ind	1A	2A	3A	4A	4B	5A	6A	7A	B**	8A	11A	B**	fus	ind	2B	2C	4C	4D	6B	8B	10A	12A	14A	B**
χ_1	+	1	1	1	1	1	1	1	1	1	1	1	1	:	++	1	1	1	1	1	1	1	1	1	1
χ_2	+	21	5	3	1	1	1	−1	0	0	−1	−1	−1	:	++	7	−1	−1	3	1	1	−1	−1	0	0
χ_3	o	45	−3	0	1	1	0	0	b7	**	−1	1	1	:	oo	3	−5	3	−1	0	1	0	0	b7	**
χ_4	o	45	−3	0	1	1	0	0	**	b7	−1	1	1	:	oo	3	−5	3	−1	0	1	0	0	**	++
χ_5	+	55	7	1	3	−1	0	1	−1	−1	1	0	0	:	++	13	5	1	1	1	−1	0	1	−1	−1
χ_6	+	99	3	0	3	−1	−1	0	1	1	−1	0	0	:	++	15	−1	3	−1	0	−1	−1	0	1	1
χ_7	+	154	10	1	−2	2	−1	1	0	0	0	0	0	:	++	14	6	2	2	−1	0	1	−1	0	0
χ_8	+	210	2	3	−2	−2	0	−1	0	0	0	1	1	:	++	14	−10	−2	2	−1	0	0	1	0	0
χ_9	+	231	7	−3	−1	−1	1	1	0	0	−1	0	0	:	++	7	−9	−1	−1	1	−1	1	−1	0	0
χ_{10}	o	280	−8	1	0	0	0	1	0	0	0	b11	**	∣	+	0	0	0	0	0	0	0	0	0	0
χ_{11}	o	280	−8	1	0	0	0	1	0	0	0	**	b11	∣											
χ_{12}	+	385	1	−2	1	1	0	−2	0	0	1	0	0	:	++	21	5	−3	−3	0	1	0	0	0	0
	ind	1	2	3	4	4	5	6	7	7	8	11	11	fus	ind	2	2	4	4	6	8	10	12	14	14
		2	2	6	4		10	6	14	14	8	22	22			2		4		6		10	12	14	14
χ_{13}	o	10	2	1	2	0	0	−1	b7	**	0	−1	−1	:	oo	4	0	−2	0	1	0	0	1	−b7	**
χ_{14}	o	10	2	1	2	0	0	−1	**	b7	0	−1	−1	:	oo	4	0	−2	0	1	0	0	1	**	−b7
χ_{15}	+	56	−8	2	0	0	1	−2	0	0	0	1	1	:	++	0	0	0	0	0	0	r5	0	0	0
χ_{16}	+	120	−8	3	0	0	0	1	1	1	0	−1	−1	:	++	8	0	4	0	−1	0	0	1	1	1
χ_{17}	o	126	6	0	−2	0	1	0	0	0	0	b11	**	∣	+	0	0	0	0	0	0	0	0	0	0
χ_{18}	o	126	6	0	−2	0	1	0	0	0	0	**	b11	∣											
χ_{19}	o	154	2	1	−2	0	−1	−1	0	0	2i	0	0	∣	+	0	0	0	0	0	0	0	0	0	0
χ_{20}	o	154	2	1	−2	0	−1	−1	0	0	−2i	0	0	∣											
χ_{21}	+	210	10	3	2	0	0	1	0	0	0	1	1	:	++	28	0	2	0	1	0	0	−1	0	0
χ_{22}	+	330	2	−3	2	0	0	−1	1	1	0	0	0	:	++	20	0	−2	0	−1	0	0	1	−1	−1
χ_{23}	+	440	−8	−1	0	0	0	1	−1	−1	0	0	0	:	++	8	0	−4	0	−1	0	0	−1	1	1
	ind	1	2	3	4	8	5	6	7	7	8	11	11	fus	ind	2	4	8	4	6	8	20	24	14	14
		4	4	12	4		20	12	28	28	8	44	44			2		8		6		20	24	14	14
		2		6			10		14	14	8	22	22												
		4		12			20		28	28	8	44	44												
χ_{24}	o2	56	0	2	0	0	1	0	0	0	2z8	1	1	∣	o2										
χ_{25}	o2	56	0	2	0	0	1	0	0	0	−2z8	1	1	*											
χ_{26}	o2	144	0	0	0	0	−1	0	−b7	**	0	1	1	*	o										
χ_{27}	o2	144	0	0	0	0	−1	0	**	−b7	0	1	1	*	o										
χ_{28}	o2	160	0	−2	0	0	0	0	−1	−1	0	−b11	**	∣	−2										
χ_{29}	o2	160	0	−2	0	0	0	0	−1	−1	0	**	−b11	∣											
χ_{30}	o2	176	0	−4	0	0	1	0	1	1	0	0	0	*	−										
χ_{31}	o2	560	0	2	0	0	0	0	0	0	0	−1	−1	*	−										

Table 10F10. Mathieu group M_{22}: Order = 443,520 = $2^7 3^2 5 \cdot 7 \cdot 11$, Mult = 12, Out = 2

		1*A*	2*A*	3*A*	4*A*	4*B*	5*A*	6*A*	7*A*	*B***	8*A*	11*A*	*B***			2*B*	2*C*	4*C*	4*D*	6*B*	8*B*	10*A*	12*A*	14*A*	*B***
	ind	1	2	3	4	4	5	6	7	7	8	11	11	fus	ind	2	2	4	4	6	8	10	12	14	14
		3	6		12	12	15	6	21	21	24	33	33												
		3	6		12	12	15	6	21	21	24	33	33												
χ_{32}	o2	21	5	0	1	1	1	2	0	0	−1	−1	−1	*	+										
χ_{33}	o2	45	−3	0	1	1	0	0	*b*7	**	−1	1	1	*	+										
χ_{34}	o2	45	−3	0	1	1	0	0	**	*b*7	−1	1	1	*	+										
χ_{35}	o2	99	3	0	3	−1	−1	0	1	1	−1	0	0	*	+										
χ_{36}	o2	105	9	0	1	1	0	0	0	0	1	−*b*11	**	↓	+2										
χ_{37}	o2	105	9	0	1	1	0	0	0	0	1	**	−*b*11	*											
χ_{38}	o2	210	2	0	−2	−2	0	2	0	0	0	1	1	*	+										
χ_{39}	o2	231	7	0	−1	−1	1	−2	0	0	−1	0	0	*	+										
χ_{40}	o2	231	−9	0	3	−1	1	0	0	0	1	0	0	*	+										
χ_{41}	o2	330	−6	0	−2	2	0	0	1	1	0	0	0	*	+										
χ_{42}	o2	384	0	0	0	0	−1	0	−1	−1	0	−1	−1	*	+										
	ind	1	2	3	4	4	5	6	7	7	8	11	11	fus	ind	2	2	4	4	6	8	10	12	14	14
		6	6	6	12	12	30	6	42	42	24	66	66			2		4		6		10	12	14	14
		3	6		12	12	15	6	21	21	24	33	33												
		2	2		4		10	6	14	14	8	22	22												
		3	6		12		15	6	21	21	24	33	33												
		6	6		12		30	6	42	42	24	66	66												
χ_{43}	o2	66	−6	0	2	0	1	0	*b*7	**	0	0	0	*	o										
χ_{44}	o2	66	−6	0	2	0	1	0	**	*b*7	0	0	0	*	o										
χ_{45}	o2	120	−8	0	0	0	0	−2	1	1	0	−1	−1	*	+										
χ_{46}	o2	126	6	0	−2	0	1	0	0	0	0	*b*11	**	↓	+2										
χ_{47}	o2	126	6	0	−2	0	1	0	0	0	0	**	*b*11	*											
χ_{48}	o2	210	10	0	2	0	0	−2	0	0	0	1	1	*	+										
χ_{49}	o2	210	−6	0	−2	0	0	0	0	0	2*i*	1	1	↓	+2										
χ_{50}	o2	210	−6	0	−2	0	0	0	0	0	−2*i*	1	1	*											
χ_{51}	o2	330	2	0	2	0	0	2	1	1	0	0	0	*	+										
χ_{52}	o2	384	0	0	0	0	−1	0	−1	−1	0	−1	−1	*	+										
	ind	1	2	3	4	8	5	6	7	7	8	11	11	fus	ind	2	4	8	4	6	8	20	24	14	14
		12	12	12	12	24	60	12	84	84	24	132	132			2		8		6		20	24	14	14
		6	6	6	12	24	30	6	42	42	24	66	66												
		4	4	12	4		20	12	28	28	8	44	44												
		3	6		12		15	6	21	21	24	33	33												
		12	12		12		60	12	84	84	24	132	132												
		2					10		14	14	8	22	22												
		12					60		84	84	24	132	132												
		3					15		21	21	24	33	33												
		4					20		28	28	8	44	44												
		6					30		42	42	24	66	66												
		12					60		84	84	24	132	132												
χ_{53}	o4	120	0	0	0	0	0	0	1	1	2*z*8	−1	−1	↓	o4										
χ_{54}	o4	120	0	0	0	0	0	0	1	1	−2*z*8	−1	−1	**											
χ_{55}	o4	144	0	0	0	0	−1	0	−*b*7	**	0	1	1	**	o2										
χ_{56}	o4	144	0	0	0	0	−1	0	**	−*b*7	0	1	1	**	o2										
χ_{57}	o4	336	0	0	0	0	1	0	0	0	0	−*b*11	**	↓	−4										
χ_{58}	o4	336	0	0	0	0	1	0	0	0	0	**	−*b*11	**											
χ_{59}	o4	384	0	0	0	0	−1	0	−1	−1	0	−1	−1	**	−2										

Table 10F11. Octad stabilizer $G(\mathcal{O})$: Order = 322560 = $2^{10}3^25.7$, Mult = 2, Out = 1

1^{24} 1^8	1^82^8 1^8	1^82^8 1^42^2	1^82^8 2^4	1^63^6 1^53	1^63^6 1^23^2	1^45^4 1^35	$1^44^42^2$ 1^42^2	$1^44^42^2$ 1^224	$1^44^42^2$ 4^2	1^37^3 17	1^37^3 17
g	21504	384	1536	180	72	15	128	16	64	14	14
1	1	1	1	1	1	1	1	1	1	1	1
7	7	3	−1	4	1	2	3	1	−1	0	0
14	14	2	6	−1	2	−1	2	0	2	0	0
15	−1	3	7	0	3	0	−1	1	3	1	1
20	20	4	4	5	−1	0	4	0	0	−1	−1
21	21	1	−3	6	0	1	1	−1	1	0	0
21	21	1	−3	−3	0	1	1	−1	1	0	0
21	21	1	−3	−3	0	1	1	−1	1	0	0
28	28	4	−4	1	1	−2	4	0	0	0	0
35	35	−5	3	5	2	0	−5	−1	−1	0	0
45	45	−3	−3	0	0	0	−3	1	1	α	$\bar{\alpha}$
45	45	−3	−3	0	0	0	−3	1	1	$\bar{\alpha}$	α
45	−3	−3	−3	0	0	0	1	1	1	α	$\bar{\alpha}$
45	−3	−3	−3	0	0	0	1	1	1	$\bar{\alpha}$	α
56	56	0	8	−4	−1	1	0	0	0	0	0
64	64	0	0	4	−2	−1	0	0	0	1	1
70	70	2	−2	−5	1	0	2	0	−2	0	0
90	−6	6	18	0	0	0	−2	0	2	−1	−1
105	−7	9	−7	0	3	0	−3	1	−3	0	0
105	−7	−3	17	0	3	0	1	−1	1	0	0
105	−7	−3	1	0	3	0	1	−1	−3	0	0
120	−8	0	8	0	−3	0	0	0	0	1	1
210	−14	6	10	0	−3	0	−2	0	−2	0	0
315	−21	3	−21	0	0	0	−1	−1	3	0	0
315	−21	−9	3	0	0	0	3	1	−1	0	0

$\alpha = \frac{1}{2}\left(-1 + i\sqrt{7}\right)$

Table 10F12. Octad stabilizer $G(\mathcal{O})$: Order = 322560 = $2^{10}3^25.7$, Mult = 2, Out = 1

8^242 1^224	$1^26^23^22^2$ 1^23^2	$1^26^23^22^2$ 12^23	$1^26^23^22^2$ 26	$1\,15\,5\,3$ 35	$1\,15\,5\,3$ 35	$1\,14\,7\,2$ 17	$1\,14\,7\,2$ 17	4^6 4^3	2^{12} 2^4	4^42^4 4^2	4^42^4 2^4	12642 26
16	24	12	12	15	15	14	14	32	512	64	384	12
1	1	1	1	1	1	1	1	1	1	1	1	1
1	1	0	−1	−1	−1	0	0	−1	−1	−1	−1	−1
0	2	−1	0	−1	−1	0	0	2	6	2	6	0
−1	−1	0	1	0	0	−1	−1	−1	−1	−1	−1	−1
0	−1	1	1	0	0	−1	−1	0	4	0	4	1
−1	0	−2	0	1	1	0	0	1	−3	1	−3	0
−1	0	1	0	β	$\bar\beta$	0	0	1	−3	1	−3	0
−1	0	1	0	$\bar\beta$	β	0	0	1	−3	1	−3	0
0	1	1	−1	1	1	0	0	0	−4	0	−4	−1
−1	2	1	0	0	0	0	0	−1	3	−1	3	0
1	0	0	0	0	0	α	$\bar\alpha$	1	−3	1	−3	0
1	0	0	0	0	0	$\bar\alpha$	α	1	−3	1	−3	0
1	0	0	0	0	0	$-\alpha$	$-\bar\alpha$	1	5	−3	−3	0
−1	0	0	0	0	0	$-\bar\alpha$	$-\alpha$	1	5	−3	−3	0
0	−1	0	−1	1	1	0	0	0	8	0	8	−1
0	−2	0	0	−1	−1	1	1	0	0	0	0	0
0	1	−1	1	0	0	0	0	−2	−2	−2	−2	1
0	0	0	0	0	0	1	1	−2	2	2	−6	0
−1	−1	0	−1	0	0	0	0	1	1	1	1	1
1	−1	0	−1	0	0	0	0	1	−7	−3	1	1
1	−1	0	1	0	0	0	0	1	9	1	−7	−1
0	1	0	−1	0	0	−1	−1	0	8	0	−8	1
0	1	0	1	0	0	0	0	2	−6	−2	2	−1
1	0	0	0	0	0	0	0	−1	3	−1	3	0
−1	0	0	0	0	0	0	0	−1	−5	3	3	0

$\beta = \frac{1}{2}\left(-1 + i\sqrt{15}\right)$

Table 10F13. Sextet stabilizer $G(\Xi)$: Order = 138240 = $2^{10}3^35$, Mult = 2, Out = 1

1^{24} 1^6	1^82^8 1^6	1^82^8 1^42	1^82^8 1^22^2	1^63^6 1^6	1^63^6 1^33	1^45^4 15	$1^44^42^2$ 1^42	$1^44^42^2$ 1^22^2	$1^44^42^2$ 1^24	1^28^242 1^24	$1^26^23^22^2$ 1^22^2	$1^26^23^22^2$ 1^33	$1^26^23^22^2$ 132	$1\,15\,5\,3$ 53	$1\,15\,5\,3$ 53
g	3072	384	768	1080	72	60	64	128	32	16	24	24	12	15	15
1	1	1	1	1	1	1	1	1	1	1	1	1	1	1	1
1	1	−1	1	1	1	1	−1	1	−1	−1	1	1	−1	1	1
5	5	3	1	5	2	0	3	1	1	1	1	2	0	0	0
5	5	−3	1	5	2	0	−3	1	−1	−1	1	2	0	0	0
5	5	1	1	5	−1	0	1	1	−1	−1	1	−1	1	0	0
5	5	−1	1	5	−1	0	−1	1	1	1	1	−1	−1	0	0
6	6	0	−2	−3	0	1	0	−2	0	0	1	0	0	β	$\bar{\beta}$
6	6	0	−2	−3	0	1	0	−2	0	0	1	0	0	$\bar{\beta}$	β
9	9	3	1	9	0	−1	3	1	−1	−1	1	0	0	−1	−1
9	9	−3	1	9	0	−1	−3	1	1	1	1	0	0	−1	−1
10	10	2	−2	10	1	0	2	−2	0	0	−2	1	−1	0	0
10	10	−2	−2	10	1	0	−2	−2	0	0	−2	1	1	0	0
12	12	0	4	−6	0	2	0	4	0	0	−2	0	0	−1	−1
16	16	0	0	16	−2	1	0	0	0	0	0	−2	0	1	1
18	2	4	6	0	3	3	0	2	2	0	0	−1	1	0	0
18	2	−4	6	0	3	3	0	2	−2	0	0	−1	−1	0	0
18	18	0	2	−9	0	−2	0	2	0	0	−1	0	0	1	1
30	30	0	−2	−15	0	0	0	−2	0	0	1	0	0	0	0
45	−3	−3	−3	0	0	0	1	1	1	−1	0	0	0	0	0
45	−3	3	−3	0	0	0	−1	1	−1	1	0	0	0	0	0
45	−3	−3	9	0	0	0	1	−3	−1	1	0	0	0	0	0
45	−3	3	9	0	0	0	−1	−3	1	−1	0	0	0	0	0
72	8	8	0	0	3	−3	0	0	0	0	0	−1	−1	0	0
72	8	−8	0	0	3	−3	0	0	0	0	0	−1	1	0	0
90	10	4	6	0	−3	0	0	2	−2	0	0	1	1	0	0
90	10	−4	6	0	−3	0	0	2	2	0	0	1	−1	0	0
90	−6	−6	6	0	0	0	2	−2	0	0	0	0	0	0	0
90	−6	6	6	0	0	0	−2	−2	0	0	0	0	0	0	0
108	12	0	−12	0	0	3	0	−4	0	0	0	0	0	0	0
135	−9	3	3	0	0	0	−1	−1	1	−1	0	0	0	0	0
135	−9	−3	3	0	0	0	1	−1	−1	1	0	0	0	0	0
135	−9	−3	−9	0	0	0	1	3	1	−1	0	0	0	0	0
135	−9	3	−9	0	0	0	−1	3	−1	1	0	0	0	0	0

$\beta = \frac{1}{2}\left(-1 + i\sqrt{15}\right)$

Table 10F15. Sextet stabilizer $G(\Xi)$: Order = 138240 = $2^{10}3^35$, Mult = 2, Out = 1

12^2 6	6^4 3^2	6^4 6	4^6 2^3	4^6 1^24	4^6 42	3^8 3^2	2^{12} 1^6	2^{12} 1^22^2	2^{12} 2^3	10^22^2 15	4^42^4 1^42	4^42^4 2^3	4^42^4 1^22^242	4^42^4 132	12642 42	12642
12	24	12	96	32	32	72	7680	256	384	20	384	128	128	96	12	12
1	1	1	1	1	1	1	1	1	1	1	1	1	1	1	1	1
−1	1	−1	−1	−1	1	1	1	1	−1	1	−1	−1	1	1	−1	1
−1	−1	−1	−1	1	−1	−1	5	1	−1	0	3	−1	1	−1	0	−1
1	−1	1	1	−1	−1	−1	5	1	1	0	−3	1	1	−1	0	−1
0	2	0	−3	−1	−1	2	5	1	−3	0	1	−3	1	−1	1	−1
0	2	0	3	1	−1	2	5	1	3	0	−1	3	1	−1	−1	−1
0	0	0	0	0	2	0	6	−2	0	1	0	0	−2	2	0	−1
0	0	0	0	0	2	0	6	−2	0	1	0	0	−2	2	0	−1
0	0	0	3	−1	1	0	9	1	3	−1	3	3	1	1	0	1
0	0	0	−3	1	1	0	9	1	−3	−1	−3	−3	1	1	0	1
1	1	1	−2	0	0	1	10	−2	−2	0	2	−2	−2	0	−1	0
−1	1	−1	2	0	0	1	10	−2	2	0	−2	2	−2	0	1	0
0	0	0	0	0	0	0	12	4	0	2	0	0	4	0	0	0
0	−2	0	0	0	0	−2	16	0	0	1	0	0	0	0	0	0
0	0	0	0	−2	0	0	−6	−2	0	−1	−4	0	−2	0	−1	0
0	0	0	0	2	0	0	−6	−2	0	−1	4	0	−2	0	1	0
0	0	0	0	0	2	0	18	2	0	−2	0	0	2	2	0	−1
0	0	0	0	0	−2	0	30	−2	0	0	0	0	−2	−2	0	1
1	−1	−1	1	1	1	3	5	5	5	0	−3	−3	−3	−3	0	0
−1	−1	1	−1	−1	1	3	5	5	−5	0	3	3	−3	−3	0	0
1	−1	−1	1	−1	−1	3	5	1	−7	0	−3	1	1	3	0	0
−1	−1	1	−1	1	−1	3	5	1	7	0	3	−1	1	3	0	0
0	0	0	0	0	0	0	−24	0	0	1	−8	0	0	0	1	0
0	0	0	0	0	0	0	−24	0	0	1	8	0	0	0	−1	0
0	0	0	0	2	0	0	−30	−2	0	0	−4	0	−2	0	−1	0
0	0	0	0	−2	0	0	−30	−2	0	0	4	0	−2	0	1	0
−1	1	1	2	0	0	−3	10	6	−2	0	−6	−2	−2	0	0	0
1	1	−1	−2	0	0	−3	10	6	2	0	6	2	−2	0	0	0
0	0	0	0	0	0	0	−36	4	0	−1	0	0	4	0	0	0
0	0	0	3	1	1	0	15	−5	−9	0	3	−1	3	−3	0	0
0	0	0	−3	−1	1	0	15	−5	9	0	−3	1	3	−3	0	0
0	0	0	−3	1	−1	0	15	−1	−3	0	−3	5	−1	3	0	0
0	0	0	3	−1	−1	0	15	−1	3	0	3	−5	−1	3	0	0

Table 10F15. Trio stabilizer $G(\mathcal{J})$: Order = $64512 = 2^{10}3^{2}7$, Mult = 2, Out = 1

1^{24}	$1^{8}2^{8}$	$1^{8}2^{8}$	$1^{8}2^{8}$	$1^{6}3^{6}$	$1^{4}4^{4}2^{2}$	$1^{4}4^{4}2^{2}$	$1^{4}4^{4}2^{2}$	$1^{3}7^{3}$	$1^{3}7^{3}$	$1^{2}8^{2}42$	$1^{2}6^{2}3^{2}2^{2}$	$1^{2}6^{2}3^{2}2^{2}$	$1\,14\,7\,2$	$1\,14\,7\,2$	12^{2}
1^{3}	1^{3}	1^{3}	$1\,2$	1^{3}	1^{3}	1^{3}	$1\,2$	1^{3}	1^{3}	$1\,2$	1^{3}	$1\,2$	$1\,2$	$1\,2$	3
1^{7}	1^{7}	$1^{3}2^{2}$	1^{7}	$1\,3^{2}$	$1^{2}2^{2}$	$1\,2\,4$	$1^{3}2^{2}$	7	7	$1\,2\,4$	$1\,3^{2}$	$1\,3^{2}$	7	7	$1\,2\,4$
g	3072	256	2688	72	128	32	64	42	42	16	24	12	14	14	12
1	1	1	1	1	1	1	1	1	1	1	1	1	1	1	1
1	1	1	−1	1	1	1	−1	1	1	−1	1	−1	−1	−1	1
2	2	2	0	2	2	2	0	2	2	0	2	0	0	0	−1
3	3	−1	3	0	−1	1	−1	α	$\bar{\alpha}$	1	0	0	α	$\bar{\alpha}$	1
3	3	−1	3	0	−1	1	−1	$\bar{\alpha}$	α	1	0	0	$\bar{\alpha}$	α	1
3	3	−1	−3	0	−1	1	1	α	$\bar{\alpha}$	−1	0	0	$-\alpha$	$-\bar{\alpha}$	1
3	3	−1	−3	0	−1	1	1	$\bar{\alpha}$	α	−1	0	0	$-\bar{\alpha}$	$-\alpha$	1
6	6	−2	0	0	−2	2	0	2α	$2\bar{\alpha}$	0	0	0	0	0	−1
6	6	−2	0	0	−2	2	0	$2\bar{\alpha}$	2α	0	0	0	0	0	−1
6	6	2	6	0	2	0	2	−1	−1	0	0	0	−1	−1	0
6	6	2	−6	0	2	0	−2	−1	−1	0	0	0	1	1	0
7	7	−1	7	1	−1	−1	−1	0	0	−1	1	1	0	0	−1
7	7	−1	−7	1	−1	−1	1	0	0	1	1	−1	0	0	−1
8	8	0	8	−1	0	0	0	1	1	0	−1	−1	1	1	0
8	8	0	−8	−1	0	0	0	1	1	0	−1	1	−1	−1	0
12	12	4	0	0	4	0	0	−2	−2	0	0	0	0	0	0
14	14	−2	0	2	−2	−2	0	0	0	0	2	0	0	0	1
16	16	0	0	−2	0	0	0	2	2	0	−2	0	0	0	0
21	5	5	7	3	1	1	3	0	0	1	−1	1	0	0	0
21	5	5	−7	3	1	1	−3	0	0	−1	−1	−1	0	0	0
21	5	1	7	3	−3	−1	−1	0	0	−1	−1	1	0	0	0
21	5	1	−7	3	−3	−1	1	0	0	1	−1	−1	0	0	0
42	−6	−2	0	0	2	0	0	0	0	0	0	0	0	0	0
42	−6	−2	0	0	2	0	0	0	0	0	0	0	0	0	0
42	10	6	14	−3	−2	0	2	0	0	0	1	−1	0	0	0
42	10	6	−14	−3	−2	0	−2	0	0	0	1	1	0	0	0
63	15	−5	21	0	−1	1	−3	0	0	1	0	0	0	0	0
63	15	−5	−21	0	−1	1	3	0	0	−1	0	0	0	0	0
63	15	−1	21	0	3	−1	1	0	0	−1	0	0	0	0	0
63	15	−1	−21	0	3	−1	−1	0	0	1	0	0	0	0	0
84	−12	−4	0	0	4	0	0	0	0	0	0	0	0	0	0
126	−18	2	0	0	−2	0	0	0	0	0	0	0	0	0	0
126	−18	2	0	0	−2	0	0	0	0	0	0	0	0	0	0

$\alpha = \frac{1}{2}\left(-1 + i\sqrt{7}\right)$

Table 10F16. Trio stabilizer $G(\mathcal{J})$: Order = 64512 = $2^{10}3^27$, Mult = 2, Out = 1

6^4	6^4	4^6	4^6	4^6	3^3	3^8	2^{12}	2^{12}	2^{12}	21 3	21 3	4^42^4	4^42^4	4^42^4	4^42^4	12 6 4 2
3	3	1^3	12	12	3	3	1^3	1^3	12	3	3	1^3	12	12	12	12
13^2	1^32^2	124	124	1^32^2	13^2	1^7	1^32^2	1^7	1^32^2	7	7	1^32^2	124	1^32^2	1^7	13^2
12	24	96	32	32	36	504	768	1536	128	21	21	128	32	128	384	12
1	1	1	1	1	1	1	1	1	1	1	1	1	1	1	1	1
1	1	1	−1	−1	1	1	1	1	−1	1	1	1	−1	−1	−1	−1
−1	−1	2	0	0	−1	−1	2	2	0	−1	−1	2	0	0	0	0
0	−1	1	1	−1	0	3	−1	3	−1	$\bar\alpha$	α	−1	1	−1	3	0
0	−1	1	1	−1	0	3	−1	3	−1	α	$\bar\alpha$	−1	1	−1	3	0
0	−1	1	−1	1	0	3	−1	3	1	$\bar\alpha$	α	−1	−1	1	−3	0
0	−1	1	−1	1	0	3	−1	3	1	α	$\bar\alpha$	−1	−1	1	−3	0
0	1	2	0	0	0	−3	−2	6	0	$-\bar\alpha$	$-\alpha$	−2	0	0	0	0
0	1	2	0	0	0	−3	−2	6	0	$-\alpha$	$-\bar\alpha$	−2	0	0	0	0
0	2	0	0	2	0	6	2	6	2	−1	−1	2	0	2	6	0
0	2	0	0	−2	0	6	2	6	−2	−1	−1	2	0	−2	−6	0
1	−1	−1	−1	−1	1	7	−1	7	−1	0	0	−1	−1	−1	7	1
1	−1	−1	1	1	1	7	−1	7	1	0	0	−1	1	1	−7	−1
−1	0	0	0	0	−1	8	0	8	0	1	1	0	0	0	8	−1
−1	0	0	0	0	−1	8	0	8	0	1	1	0	0	0	−8	1
0	−2	0	0	0	0	−6	4	12	0	1	1	4	0	0	0	0
−1	1	−2	0	0	−1	−7	−2	14	0	0	0	−2	0	0	0	0
1	0	0	0	0	1	−8	0	16	0	−1	−1	0	0	0	0	0
0	0	−3	−1	−1	0	0	−3	−3	−1	0	0	−3	−1	−1	−1	−1
0	0	−3	1	1	0	0	−3	−3	1	0	0	−3	1	1	1	1
0	0	3	1	−1	0	0	9	−3	3	0	0	1	1	3	−1	−1
0	0	3	−1	1	0	0	9	−3	−3	0	0	1	−1	−3	1	1
−1	0	0	2	0	−3	0	6	2	4	0	0	−2	−2	−4	0	0
−1	0	0	−2	0	−3	0	6	2	−4	0	0	−2	2	4	0	0
0	0	0	0	−2	0	0	6	−6	2	0	0	−2	0	2	−2	1
0	0	0	0	2	0	0	6	−6	−2	0	0	−2	0	−2	2	−1
0	0	−3	−1	1	0	0	3	−9	1	0	0	3	−1	1	−3	0
0	0	−3	1	−1	0	0	3	−9	−1	0	0	3	1	−1	3	0
0	0	3	1	1	0	0	−9	−9	−3	0	0	−1	1	−3	−3	0
0	0	3	−1	−1	0	0	−9	−9	3	0	0	−1	−1	3	3	0
1	0	0	0	0	3	0	12	4	0	0	0	−4	0	0	0	0
0	0	0	−2	0	0	0	−6	0	4	0	0	2	2	−4	0	0
0	0	0	2	0	0	0	−6	0	−4	0	0	2	−2	4	0	0

Table 10F17. Mathieu group M_{23}: Order = 10,200,960, Mult = 1, Out = 1

	;	@	@	@	@	@	@	@	@	@	@	@	@	@	@	@	@	@
			2															
		10200960	688	180	32	15	12	14	14	8	11	11	14	14	15	15	23	23
	p power		A	A	A	A	AA	A	A	A	A	A	AA	BA	AA	AA	A	A
	p′ part		A	A	A	A	AA	A	A	A	A	A	AA	BA	AA	AA	A	A
	ind	1A	2A	3A	4A	5A	6A	7A	B**	8A	11A	B**	14A	B**	15A	B**	23A	B**
χ_1	+	1	1	1	1	1	1	1	1	1	1	1	1	1	1	1	1	1
χ_2	+	22	6	4	2	2	0	1	1	0	0	0	−1	−1	−1	−1	−1	−1
χ_3	∘	45	−3	0	1	0	0	b7	**	−1	1	1	−b7	**	0	0	−1	−1
χ_4	∘	45	−3	0	1	0	0	**	b7	−1	1	1	**	−b7	0	0	−1	−1
χ_5	+	230	22	5	2	0	1	−1	−1	0	−1	−1	1	1	0	0	0	0
χ_6	+	231	7	6	−1	1	−2	0	0	−1	0	0	0	0	1	1	1	1
χ_7	∘	231	7	−3	−1	1	1	0	0	−1	0	0	0	0	b15	**	1	1
χ_8	∘	231	7	−3	−1	1	1	0	0	−1	0	0	0	0	**	b15	1	1
χ_9	+	253	13	1	1	−2	1	1	1	−1	0	0	−1	−1	1	1	0	0
χ_{10}	∘	770	−14	5	−2	0	1	0	0	0	0	0	0	0	0	0	b23	**
χ_{11}	∘	770	−14	5	−2	0	1	0	0	0	0	0	0	0	0	0	**	b23
χ_{12}	∘	896	0	−4	0	1	0	0	0	0	b11	**	0	0	1	1	−1	−1
χ_{13}	∘	896	0	−4	0	1	0	0	0	0	**	b11	0	0	1	1	−1	−1
χ_{14}	∘	990	−18	0	2	0	0	b7	**	0	0	0	b7	**	0	0	1	1
χ_{15}	∘	990	−18	0	2	0	0	**	b7	0	0	0	**	b7	0	0	1	1
χ_{16}	+	1035	27	0	−1	0	0	−1	−1	1	1	1	−1	−1	0	0	0	0
χ_{17}	+	2024	8	−1	0	−1	−1	1	1	0	0	0	1	1	−1	−1	0	0

Table 10F18. Mathieu group M_{24}: Order = 244,823,040 = $2^{10}3^{3}5.7.11.23$, Mult = 1, Out = 1

M_{24}:	;	@	@	@	@	@	@	@	@	@	@	@	@	@	@	@	@	@	@	@	@	@	@	@	@	@	@	
		244823040	21504	7680	1080	504	384	128	96	60	24	24	42	42	16	20	11	12	12	14	14	15	15	21	21	23	23	
	Cycle shape	1^{24}	$1^8 2^8$	2^{12}	$1^6 3^6$	3^8	$2^4 4^4$	$1^2 2^2 4^4$	4^6	$1^4 5^4$	$1^2 2^2 3^2 6^2$	6^4	$1^3 7^3$	$1^3 7^3$	$1^2 2\,4\,8^2$	$2^2 10^2$	$1^2 11^2$	$2\,4\,6\,12$	12^2	$1\,2\,7\,14$	$1\,2\,7\,14$	$1\,3\,5\,15$	$1\,3\,5\,15$	$3\,21$	$3\,21$	$1\,23$	$1\,23$	
	p power		A	A	A	A	A	A	B	A	AA	BB	A	A	B	AB	A	AA	BC	AA	BA	AA	AA	BB	AB	A	A	
	p′ part		A	A	A	A	A	A	B	A	AA	BB	A	A	B	AB	A	AA	BC	AA	BA	AA	AA	BB	AB	A	A	
	ind	1A	2A	2B	3A	3B	4A	4B	4C	5A	6A	6B	7A	B**	8A	10A	11A	12A	12B	14A	B**	15A	B**	21A	B**	23A	B**	
χ_1	+	1	1	1	1	1	1	1	1	1	1	1	1	1	1	1	1	1	1	1	1	1	1	1	1	1	1	χ_1
χ_2	+	23	7	−1	5	−1	−1	3	−1	3	1	−1	2	2	1	−1	1	−1	−1	0	0	0	0	−1	−1	0	0	χ_2
χ_3	o	45	−3	5	0	3	−3	1	1	0	0	−1	b7	**	−1	0	1	0	1	−b7	**	0	0	b7	**	−1	−1	χ_3
χ_4	o	45	−3	5	0	3	−3	1	1	0	0	−1	**	b7	−1	0	1	0	1	**	−b7	0	0	**	b7	−1	−1	χ_4
χ_5	o	231	7	−9	−3	0	−1	−1	3	1	1	0	0	0	−1	1	0	−1	0	0	0	b15	**	0	0	1	1	χ_5
χ_6	o	231	7	−9	−3	0	−1	−1	3	1	1	0	0	0	−1	1	0	−1	0	0	0	**	b15	0	0	1	1	χ_6
χ_7	+	252	28	12	9	0	4	4	0	2	1	0	0	0	0	2	−1	1	0	0	0	−1	−1	0	0	−1	−1	χ_7
χ_8	+	253	13	−11	10	1	−3	1	1	3	−2	1	1	1	−1	−1	0	0	1	−1	−1	0	0	1	1	0	0	χ_8
χ_9	+	483	35	3	6	0	3	3	3	−2	2	0	0	0	−1	−2	−1	0	0	0	0	1	1	0	0	0	0	χ_9
χ_{10}	o	770	−14	10	5	−7	2	−2	−2	0	1	1	0	0	0	0	0	−1	1	0	0	0	0	0	0	b23	**	χ_{10}
χ_{11}	o	770	−14	10	5	−7	2	−2	−2	0	1	1	0	0	0	0	0	−1	1	0	0	0	0	0	0	**	b23	χ_{11}
χ_{12}	o	990	−18	−10	0	3	6	2	−2	0	0	−1	b7	**	0	0	0	0	1	b7	**	0	0	b7	**	1	1	χ_{12}
χ_{13}	o	990	−18	−10	0	3	6	2	−2	0	0	−1	**	b7	0	0	0	0	1	**	b7	0	0	**	b7	1	1	χ_{13}
χ_{14}	+	1035	27	35	0	6	3	−1	3	0	0	2	−1	−1	1	0	1	0	0	−1	−1	0	0	−1	−1	0	0	χ_{14}
χ_{15}	o	1035	−21	−5	0	−3	3	3	−1	0	0	1	2b7	**	−1	0	1	0	−1	0	0	0	0	−b7	**	0	0	χ_{15}
χ_{16}	o	1035	−21	−5	0	−3	3	3	−1	0	0	1	**	2b7	−1	0	1	0	−1	0	0	0	0	**	−b7	0	0	χ_{16}
χ_{17}	+	1265	49	−15	5	8	−7	1	−3	0	1	0	−2	−2	1	0	0	−1	0	0	0	0	0	1	1	0	0	χ_{17}
χ_{18}	+	1771	−21	11	16	7	3	−5	−1	1	0	−1	0	0	−1	1	0	0	−1	0	0	1	1	0	0	0	0	χ_{18}
χ_{19}	+	2024	8	24	−1	8	8	0	0	−1	−1	0	1	1	0	−1	0	−1	0	1	1	−1	−1	1	1	0	0	χ_{19}
χ_{20}	+	2277	21	−19	0	6	−3	1	−3	−3	0	2	2	2	−1	1	0	0	0	0	0	0	0	−1	−1	0	0	χ_{20}
χ_{21}	+	3312	48	16	0	−6	0	0	0	−3	0	−2	1	1	0	1	1	0	0	−1	−1	0	0	1	1	0	0	χ_{21}
χ_{22}	+	3520	64	0	10	−8	0	0	0	0	−2	0	−1	−1	0	0	0	0	0	1	1	0	0	−1	−1	1	1	χ_{22}
χ_{23}	+	5313	49	9	−15	0	1	−3	−3	3	1	0	0	0	−1	−1	0	1	0	0	0	0	0	0	0	0	0	χ_{23}
χ_{24}	+	5544	−56	24	9	0	−8	0	0	−1	1	0	0	0	0	−1	0	1	0	0	0	−1	−1	0	0	1	1	χ_{24}
χ_{25}	+	5796	−28	36	−9	0	−4	4	0	1	−1	0	0	0	0	1	−1	−1	0	0	0	1	1	0	0	0	0	χ_{25}
χ_{26}	+	10395	−21	−45	0	0	3	−1	3	0	0	0	0	0	1	0	0	0	0	0	0	0	0	0	0	−1	−1	χ_{26}

Chapter 11. Generation Three of the Happy Family and the Pariahs

We now sketch the discoveries and constructions of the remaining fourteen sporadic simple groups, and make some comments on their properties.

The third generation of the **Happy Family** [Gr82] consists of eight simple groups (11.9) which are involved in the largest sporadic simple group, the Monster, denoted $\mathbb{M}$. The construction of $\mathbb{M}$ depends on construction and analysis of a certain commutative, nonassociative algebra, B, of dimension 196884 over the rational field. This algebra plays the auxiliary role of the Golay code in the first generation and the Leech lattice in the second [Gr82].

The six sporadic groups not members of the Happy Family are called **the Pariahs** [Gr82]. Their stories will be summarized in (11.12)–(11.17).

(11.1) Notation. We use the notations for the Leech lattice Λ, as in Chapters 9 and 10. We let $Q \cong 2_+^{1+24}$ be an extraspecial group, T the irreducible 2^{12} dimensional module, written over the rational field. There is a group H containing Q as a normal subgroup such that $C_H(Q) = Z(Q)$ and $H/Q \cong \Omega^+(24,2)$ and such that the representation $Q \to GL(T)$ extends to a representation of H (H is called a *holomorph* of Q; see (2.21) and (2.22)).

(11.2) Lemma. (i) *The quadratic form on $\Lambda/2\Lambda$ is nondegenerate and has maximal Witt index (i.e., is of "plus type");*

(ii) *There is an isomorphism of elementary abelian groups $\Lambda/2\Lambda \cong Q/Q'$ which is also an isometry in the sense that $x + 2\Lambda$ corresponds to a coset yQ' such that $\frac{1}{2}(x,x)(mod\, 2)$ is 0, 1, respectively, iff y^2 is 1 or the nontrivial generator of Q'.*

Proof. Since 13 divides the order of $Aut(\Lambda)$, we deduce that the index is maximal; see (2.24.1). Another argument is that we have already displayed a maximal totally singular subspace of $\Lambda/2\Lambda$, e.g., (10.18). (ii) This follows from (i) since Q has plus type. □

(11.3) Definition (The group C.) We have an orthogonal action of $Aut(\Lambda)$ on $\Lambda/2\Lambda$, with kernel $\{\pm 1\}$, so there is a group C_∞ between Q and H which corresponds to the image of this representation. The group $\widehat{C}$ is defined as the pullback or fiber product in the following diagram:

$$\begin{array}{ccc} \widehat{C} & \rightarrow & C_\infty \\ \downarrow & & \downarrow \\ Aut(\Lambda) & \rightarrow & Aut(\Lambda)/\{\pm 1\} \end{array} .$$

Equivalently, we may take $\widehat{C}$ to be the set of pairs $(x, y) \in Aut(\Lambda) \times C_\infty$ such that the images of x and y in $Aut(\Lambda)/\{\pm 1\}$ are equal. The centers of the two factors have order 2, if Z denotes the "diagonal" subgroup of order 2, we define $C := \widehat{C}/Z$.

The group C contains a copy of Q as a normal subgroup, but the representation of Q on T does not extend to a representation of C on Q; the above diagram shows that we have a projective representation of C on T.

(11.4) (The C-module B). We define a vector space B with the structure of a C-module. Let E be the rational vector spaces containing Λ and let $U := S^2(E)$ (symmetric tensors of degree 2), $W := E \otimes T$. Let X be the set of all cosets of Q' in Q which correspond to vectors of type 2 by an isometry as in (2.4.ii). Then $|X| = 196560/2 = 98280$. We let V be the irreducible C-module which is monomial respect to basis X for which the kernel of the action is Q'.

$$B = U \oplus V \oplus W .$$

There is a nontrivial C-invariant bilinear form on each summand. Notice that U has a 1-space I corresponding to the invariant bilinear form on E; I is a C-submodule. Define $U_0 := I^\perp \cap U$.

(11.5) (The C-algebra B). We give only a sketch. The subspace U is a subalgebra, isomorphic to the Jordan algebra of symmetric degree 24 matrices. The space V is stable under multiplication by U, with each 1-space spanned by an element of Y (see (11.4)) a U-submodule; also $U+V$ is a subalgebra with $V^2 = U+V$. Finally, W is stable under $U+V$ and $W^2 = U+V$. The product on B is C-invariant, so C acts as a group of algebra automorphisms.

(11.6) Improvements on the Construction. The original definition of the algebra structure in [Gr82] was complicated, due mainly to sign problems. For this product, the codimension 1 submodule $U_0 + V + W$ was an ideal; however, to make the product on U isomorphic to the Jordan algebra of degree 24 symmetric matrices, the product on B can be changed in a way that the automorphism group is the same. Existence of an irreducible representation of the monster of degree 196883 was predicted in 1974 by Griess [Gr75]. Norton proved around 1976 that a nontrivial commutative, nonassociative algebras structure on such a module exists, though his method gave no description (he also made assumptions about the conjugacy classes). The possible structures of such an algebra and the conditions for its having an automorphism outside C were studied simultaneously by surveying the actions of many subgroups of C on the space B; the result was the system of structure constants in [Gr82], Table 6.1 and the formula for an extra automorphism in [Gr82], Table 10.2. Proof that the linear automorphism so defined preserves the algebra structure was the hardest part of the construction [Gr82].

Many improvements on [Gr82] were made by Tits [Tits83a, 83b, 83c, 84] who showed that some definitions of [Gr82] based on guesswork may be based on a more thorough analysis. A new style construction was made by Conway [Co84] [CoSl] who used a Moufang loop to finesse the nasty sign problems so prominent in the original version. This loop (a nonassociative group) has order 2^{13} and is a kind of 2-cover of the binary Golay code; its creation was an idea of Richard Parker; see [Gr86a], which defines and gives the foundation of the class of loops called **code loops**, of which Parker's loop is an example. Conway's presentation is summarized in [Atlas]. We make no attempt to survey this theory here, but do refer to the development of the loop theory in the theory of p-locals in sporadic groups and Lie groups; see [Gr86a, 87a, 88, 90, 93] [Rich91, 92, 94, 95] [Jo] [Buri] [Hsu].

The algebra B is not a classic nonassociative algebra. An algebra of dimension n satisfies a nontrivial polynomial identity of degree at most $n+1$; B satisfies no nontrivial identity in commuting variables of degree less than 6 [Gr86b]. For some related algebras, see [Smi].

In [Gr82], the subgroup of $Aut(B)$ generated by C and the particular extra automorphism was identified as a simple group of the right order, thus proving existence of a simple group of the right order and local properties. The full automorphism group of a finite dimensional algebra is an algebraic group. In [Tits84], Tits showed that $Aut(B)$ was exactly the monster. In [Co84], Conway gave a short argument with idempotents in B that $Aut(B)$ is finite and in [Tits84], Tits identifies the centralizer of an involution in $Aut(B)$ as C (not a larger group). The proof that the group order is right involves quoting harder theorems; see Chapter 13 of [Gr82]. In 1988, a uniqueness proof for the monster was given by Griess, Meierfrankenfeld and Segev [GMS]; see (11.7).

(11.7) Uniqueness Results for the Monster. First, we define a group of *Monster type* to be a finite group G containing a pair of involutions z, t such that (i) $C(z) \cong 2_+^{1+24}Co_1$; and (ii) $C(t)$ is a double cover of Fischer's $\{3,4\}$-transposition group, discovered in 1973 [Fi73] (this is the group later called the "baby monster"). It is trivial that such a group is simple. See [GMS] for a fuller discussion of the hypotheses and background.

Theorem (Uniqueness of the Monster). *A group of monster type is unique up to isomorphism* [GMS].

We define a graph Γ on the conjugacy class of t in G (the class is usually called $2A$) by letting the conjugacy class be the vertex set and by connecting two vertices with an edge if their product is in the class. Clearly, G is contained in the automorphism group of Γ.

Theorem. ([GMS], (5.10)) $Aut(\Gamma) = G$.

It is worth mentioning that G contains a unique conjugacy class of fours groups which are $2A$-pure (its normalizer has shape $2^2 {\cdot} {}^2E_6(2){:}\Sigma_3$) but no eights groups which are $2A$-pure (3.3) [GMS].

(11.9) Consequences. Since C is a subgroup of $G := Aut(B)$, G involves all twelve sporadic simple groups constructed in Chapters 5-10. From existence of G, we deduce existence of eight additional sporadic simple groups. These are listed below; details may be found in Chapter 13 of [Gr82].

Table 11.1. Third generation of the Happy Family

Group	Discoverer (date)	First construction (date)	Order
He	D. Held (1968)	G. Higman, J. McKay (1968?)	$2^{10}3^3 5^2 7^3.17$
Fi_{22}	B. Fischer (1968)	B. Fischer (1969)	$2^{17}3^9 5^2 7.11.13$
Fi_{23}	B. Fischer (1968)	B. Fischer (1969)	$2^{18}3^{13}5^2 7.11.13.17.23$
Fi'_{24}	B. Fischer (1968)	B. Fischer (1969)	$2^{21}3^{16}5^2 7^3 11.13.17.23.29$
F_2	B. Fischer (1973)	J. Leon, C. Sims (1975?)	$2^{41}3^{13}5^6 7^2 11.13.17.19.23.31.47$
$F_1 = IM$	B. Fischer, R. Griess (1973)	R. Griess (14 January, 1980)	$2^{46}3^{20}5^9 7^6 11^2 13^3 17.19.23.29.31.41.47.59.71$
F_3	J. Thompson (1973)	P. Smith (1974)	$2^{15}3^{10}5^3 7^2 13.19.31$
F_5	K. Harada (1973)	S. Norton (1974)	$2^{14}3^6 5^6 7.11.19$

(11.10) Remarks. The existence corollary (11.9) for these eight simple groups replaces earlier existence proofs. The three Fischer groups were proven to exist with Fischer's theory of 3-transposition groups [Fi69, 71a]. The group of Held was constructed by Graham Higman and John McKay using a computer, as a permutation group with point stabilizer $Sp(4,4){:}2$ [HM]. The group F_3 of Thompson was constructed by him and Peter Smith with computer [Th76], as a linear group of degree 248 which preserves and integral lattice; this group acts absolutely irreducibly on this lattice modulo every prime. The group F_5 of Harada was constructed by Norton with computer [No75]. The baby monster was constructed as a permutation group by Leon and Sims with computer [LS]. It is interesting that all sporadic groups involved in the monster occur in local subgroups (except for the monster itself). The corresponding statement for the first two generations of the Happy Family would not be correct, e.g. Co_2 in Co_1 and M_{23} in M_{24}. Also, it is worth noting that some irreducible degrees for sporadic groups are the dimensions of Lie algebras; we have degree 78 for Fi_{22}, 133 for F_5 and 248 for F_3; these groups are not embedded in the suggested complex Lie groups but we do have embeddings in positive characteristic: Fi_{22} in ${}^2E_6(2)$ [Fi71b] and F_3 in $E_8(3)$ [Th76]. There are similar examples involving other simple groups, e.g. $G_2(3)$ has a degree 14 character yet is not contained in the complex group of type G_2 [CoGr].

(11.11) Moonshine. The term "Moonshine" is hard to make precise. In general terms, it means connections between finite simple groups and other areas of mathematics and mathematical physics which are surprising because they span significant gaps and involve nontrivial information in two or more areas. The "original moonshine" is the particular theory of McKay, Thompson, Conway and Norton [CN] [Broué] [Gr86b] [Th79b] which connects the conjugacy classes of the monster with a set of genus zero function fields. It began with McKay's obervation that the elliptic modular function $j = q^{-1} + 744 + 196884q + \cdots$ has coefficient $196884 = 1 + 196883$, the sum of two irreducible degrees of the Monster; Thompson noted that higher coefficients also seem to have similar expressions and he suggested that this may be explained by an infinite dimensional graded representation of the monster [Th79b]. Shortly thereafter, a fuller connection between conjugacy classes of the Monster and modular forms of genus zero was developed by Conway and Norton [CN]. See the exposition [Broué]. Recently [Bor] gave a proof of formulas which were conjectured in [CN]. Other coincidences involving the monster are considered part of moonshine, for instance the following observation of McKay: if we fix an involution t in conjugacy class $2A$ and choose a representative u from each of the 9 orbits of $C(t)$ on $2A$, then the conjugacy class of tu is $1A, 2A, 2B, 3A, 3C, 4A, 4B, 5A, 6A$ [GMS]; the orders of tu form the coefficients the highest root in E_8 with respect to a base of the root system. There is not yet an explanation for this provocative observation. See [Gr86b] for a survey of moonshine-related topics and articles by Geoffrey Mason [Ma84, 85].

The Monster also plays a role in infinite dimensional representation theory associated to conformal field theory and vertex operator algebras [FLM], where a moonshine module was constructed. The 196884-dimensional algebra B has taken on quite an importance in these fields, and has become known as "the Griess algebra". It is amusing that in recent years, I have heard the algebra referred to more often than the group.

Now to describe **the Pariahs**, the six sporadic groups outside the Happy Family. For proofs that five of them are not involved in the Monster, see [Gr82] and, for J_1, see [Wil86]. In the cases of the Lyons group and the fourth Janko group, this is obvious from Lagrange's theorem.

(11.12) The First Janko Group, J_1 (1965). It was discovered by Janko as a solution to a centralizer of involution problem [J]. It is the unique simple group which has abelian Sylow 2-groups and contains an involution with centralizer isomorphic to $2 \times PSL(2,5) \cong 2 \times Alt_5$, constructed with computer by John Cannon as a subgroup of $G_2(11)$ generated by a pair of degree 7 matrices over the integers modulo 11.

It has been realized as a permutation group of degree 266 on the cosets of a $PSL(2,11)$-subgroup [Co71]. The order is $2^3.3.5.7.11.19 = 19.20.21.22 = 55.56.57 = 175560$. The order is also writeable as the "polynomial" $11(11+1)(11^3-1)$, which suggested that J_1 was part of a series of groups over finite fields of characteristic 11; nothing came of this.

(11.13) The Third Janko Group, J_3 (1968). Again, this is a group discovered by Janko as a solution to a centralizer of involution problem. The involution centralizer is $2_-^{1+4}{:}Alt_5$, split extension (therefore not the standard holomorph, which embeds in $GL(4,\mathbf{C})$). It turns out that two finite simple groups have such a centralizer of involution: one is the group of Hall-Janko (notated HJ or J_2), described in (10.38.3), and the other is J_3 of order $2^7 3^5 5 \cdot 17 \cdot 19$. The latter was constructed by G. Higman and J. McKay with computer [HMcK]. A construction of J_3 as a group of automorphisms of a 170-dimensional algebra was made by Frohardt [Froh].

(11.14) The Lyons Group, Ly (1969). This is a group discovered by a centralizer of involution characterization [Ly]. Lyons considered finite simple groups containing an involution whose centralizer is isomorphic to $2 \cdot A_{11}$, the covering group of the alternating group of degree 11. Such a group has a single class of involutions and order $2^8 3^5 5.7.11.37.61.67$. The lowest degree faithful permutation representation seems to be about 9 million, of large rank, on the cosets of a $G_2(5)$-subgroup. Such a permutation group was constructed by Sims, using computer [Sims72]. Among its interesting subgroups are these:

a triple cover of the simple group of McLaughlin, extended by a group of outer automorphisms of order 2 (we remark that the simple group of McLaughlin has an involution centralizer of shape $2 \cdot Alt_8$);

a 3-local of the form $3^5 : [2 \times M_{11}]$;

a 5-local of the form $5^3 \cdot SL(3,5)$ (nonsplit; see (2.17)); this group is said to be in F_2 [Atlas].

(11.15) The Rudvalis Group, Ru (1972). This group was discovered as a rank 3 permutation group; see Appendix 10A. Rudvalis tried to find new simple groups which were rank 3 groups with likely candidates for a point stablizer. Let G be a rank 3 group and H a point stabilizer in G. Rudvalis tried $H = {}^2F_4(2)$, the Ree group over the field of 2 elements (it is not simple, as are the other Ree groups ${}^2F_4(2^{2n+1})$ for $n \geq 1$, but has a simple subgroup of index 2, as shown by Tits [Tits64]). Rudvalis tried to exploit the arithmetic relations (10A.5), with one 2-point stabilizer as the rank 1 parabolic subgroup of order $2^{12}5$ and index 1755. He found that the arithmetic conditions were satisfied if there were a subgroup of index 2304, and conjectured that such a subgroup exists and is isomorphic to $PSL(2,25) \cdot 2$, the group of even permutations of $P\Gamma L(2,25)$ on the 26 points of the projective line over the field of 25 elements; the parameters had to be $n = 4060, k = 1755, \ell = 2304, \mu = 780, \lambda = 732$ (10A.8); [Ru]. The group was constructed shortly afterwards by Conway and Wales who defined a degree 28 matrix group over the Gaussian numbers which is a double cover of a simple group with all the above properties.

There are two classes of involutions, with centralizers of shapes $2^{1+4+6}.Sym_5$ and $[2^2 \times Sz(8)]{:}3$.

Local subgroups include $3 \cdot M_{10}{:}2$, $2^{3+8}{:}GL(3,2)$; the Sylow 3-group has the shape 3^{1+2} and there is just one class of elements of order 3 (in fact this is already true in the subgroup ${}^2F_4(2)$).

$2 \cdot Ru$ embeds in $E_7(5)$, and on its 56-dimensional module, the action of $2 \cdot Ru$ has a pair of 28-dimensional irreducibles [GrRy].

(11.16) The O'Nan Group, $O'N$ (1973). The path to discovery was rather unusual. O'Nan was studying finite simple groups G with the following property: for every elementary abelian 2-subgroup, $E \leq G$, $N_E(E)/C_G(E) \cong GL(E)$. He was led to a case where G contains a subgroup $E \cong 2^3$ such that $N_G(E) \cong 4^3 \cdot GL(3,2)$ and an involution whose centralizer has the form $4.PSL(3,4).2$. He then determined its order and many of its properties [O'N]. The group was constructed by Sims [Sims0]. Such a group has an outer automorphism of order 2 whose fixed point subgroup is isomorphic to J_1. See [Gr87a] for a discussion of cohomology issues connected to $4^3 \cdot GL(3,2)$ and similar 2-locals in simple groups.

(11.17) The Fourth Janko Group, J_4 (1975). This group was discovered, in good Janko tradition, by a centralizer of involution problem. The relevant centralizer had the form $2_+^{1+12} \cdot 3 \cdot M_{22}{:}2$; the centralizer is nonsplit over O_2 and contains a perfect group $6 \cdot M_{22}$. Such a simple group has order $2^{21}3^3 5.7.11^3.23.29.31.37.43$. also contains subgroups of the form $2^{11}{:}M_{24}$, with the module isomorphic to the even part of the Golay cocode (a nonsplit version of this appears in Fi_{24}), a copy of $2^{10}{:}GL(5,2)$ and a group of the form $2^{3+12} \cdot [Sym_5 \times GL(3,2)]$. A group with such properties was constructed as a subgroup of $GL(112,2)$ by Benson, Conway, Norton, Parker and Thackray [No80], with computer. The smallest degree of a faithful representation in characteristic 0 is 1333. Note that the title of [Ja76] is off since the Schur multiplier of M_{22} is cyclic of order 12, not 6; the error was noticed and the correction was made after [Ja76] was published, in the late 1970s by Fong, [Gr80a] and [Ma79, 82].

(11.18) Involvement of Sporadic Groups in Sporadic Groups. This table originated as Table 14.1 [Gr82]; a later version, resolving most question marks, appeared in [Atlas], p. 238; see also [Wil86].

(11.19) The Program to Classify Finite Simple Groups. In the mid 1970s, there were increased developments in many areas of the classification of finite simple groups. While I shall not attempt a summary of this story, I do want do record a general impression among the simple group theorists which seems to me to be historically significant. (I do so at the risk of unfairly emphasising one part of the classification story). About 1977, Gorenstein and Lyons announced their important "Tricotomy Theorem", which focused the conclusion of the classification of simple groups on three specific major problems concerning the groups of characteristic two type. Because these problems were so specific, this was considered the first indication that the program to classify finite simple groups might realistically achieve closure (a contemporary summary of this point, with some de-emphasis on its dramatic impact, is given in the bottom half of page 40 of [GLS]). Of course, many group theorists in that era still hoped to find more simple groups; indeed Janko had "done it again" in spring 1975. So, there was an aura of good-natured conflict among the finite group theorists (sometimes within the same group theorist!): the wish for more of these elegant simple groups and the wish to settle the classification.

The classification of finite simple groups was announced by Gorenstein in January of 1981; see [Go82]. This was a bit optimistic since proofs of some outstanding cases existed only in manuscript form and not every one of these was submitted promptly. Despite this delay in resolution, most finite group theorists continue to believe that the program to classify the simple groups is sound. At the present time, efforts of Lyons and Solomon ([GLS], and upcoming sequels) are actively bringing about closure on this important classification; see the recent survey [Sol].

Efforts to find more simple groups certainly continued after 1975, but no one has since presented a serious candidate for a 27th sporadic. Hope that there are more of these beautiful and fascinating creatures nearly disappeared by the early 80s. We have to face the fact that we do not know a single theme which convincingly unites these 26 sporadic groups.

Appendix. Some Comments on the Atlas

My preferred symbols for the finite simple groups differ somewhat with those in [Atlas]; see the Introduction of this book. My notations for group extensions given in Chapter 2 are generally compatible with those of [Atlas], page xx. However, I do feel that a subgroup of index 2 in $A \times B$ should be notated $[A \times B]\frac{1}{2}$ and not $\frac{1}{2}[A \times B]$ as in the Atlas (page xx, column 2); the reason is that the group extension notation reads left-to-right in order of ascending factors, so dropping down to a subgroup of index 2 corresponds to removing a factor of order 2 *from the top*.

The Atlas of Finite Groups [Atlas] was published in 1985 and the Atlas of Brauer Characters [AtlasBC] was published in 1995. The latter contains some corrections and additions to the former. I want to make a few remarks about the attributions (or lack of them) made in these books.

The table of involvements of sporadic groups is a slight update of a table which is originally due to me; see [Gr82], Table 14.1 and (11.18) in this book. I feel that principal credit for such a table should be given to me.

The foundations of the theory of code loops was established in [Gr86a], which clearly credits the ideas and initial applications to Richard Parker and John Conway. I feel that my foundation should have been referred to in [Atlas] where loops were used, e.g. in the sections for Fi_{24} and the monster, $\mathbb{M} = F_1$. The preprint of Code Loops [Gr86a] was circulated in 1984, though it was not referred to in [Atlas], nor was [Gr86a] referred to in [AtlasBC]. There are further interesting developements in loops and their use in constructing sporadic p-locals and exotic p-locals in Lie groups. See the discussion in (11.6).

References

[AlpGor] Jon Alperin and Daniel Gorenstein: A Vanishing Theorem for Cohomology. Proc. Amer. Math. Soc. **32** (1972) 87–88.

[Artin] Emil Artin: Geometric Algebra. Interscience, New York 1957.

[Atlas] Conway, Curtis, Norton, Parker, Wilson: An Atlas of finite groups. Clarendon Press, Oxford 1985.

[AtlasBC] Christoph Jansen, Klaus Lux, Richard Parker, Robert Wilson: An Atlas of Brauer Characters, Oxford Science Publications. Clarendon Press, Oxford 1995.

[Ben] David Benson: Modular Representation Theory: New Trends and Methods. Lecture Notes in Mathematics, Vol. 1081, Springer, Berlin 1984.

[Bl] Norman Blackburn: The extension theory of the symmetric and alternating groups. Math. Zeitschrift **117** (1970) 191–206.

[Bor] R. Borcherds: Monstrous moonshine and monstrous Lie superalgebras. Inventiones mathematicae 109, (1992) 405–444.

[Bour] Nicolas Bourbaki: Groupes et Algèbres de Lie; Chapitres 4, 5, 6. Masson, Paris 1981.

[BrSah], Richard Brauer and Chih-Han Sah: Theory of Finite Groups. W. A. Benjamin, Inc., 1969.

[Broué] Michel Broué: Groupes finis, series formelles, formes modulaire, article in Seminaire sur les Groupes Finis, Tome I, Publications Mathématiques de l'Universiteé Paris VII (ouvrage Collectif: E.R.A. 944 du C.N.R.S.).

[Bur] W. Burnside: Theory of Groups of Finite Order, 2nd edition, 1911, Cambridge University Press; reprinted 1955, Dover, New York.

[Buri] Vladimir Burichenko: On a special loop, Dixon form and a lattice related to $O_7(3)$. Matematicheskii Sbornik **182**, N 10 (1991) 1408–1429. (in Russian)

[CE] Cartan, H. and Eilenberg, S.: Homological Algebra. Princeton University Press, Princeton, N.J., 1956.

[Car] Roger Carter: Simple Groups of Lie Type. Wiley-Interscience, New York 1972.

[Chev54] Claude Chevalley: The Algebraic Theory of Spinors. Columbia University Press, Morningside Heights 1954.

[Chev55] Chevalley, Claude: Sur Certains Groups Simples. Tohoku Math. J. **7** (1955) 14–66.

[Choi0] Chang Choi: Thesis, University of Michigan 1968.

[Choi72] On subgroups of M_{24}; I. Stabilizers of subsets. Trans. Amer. Math. Soc. **167** (1972) 1–27; On subgroups of M_{24}; II. The maximal subgroups. Trans. Amer. Math. Soc. **167** (1972) 29–47.

[CoGr] Arjeh Cohen and Robert L. Griess, Jr.: On simple subgroups of the complex Lie group of type E_8. Proc. Symp. Pure Math. **47** (1987) 367–405.

[Con69a] Conway, John: A Group of Order 8, 315, 553, 613, 086, 720, 000. Bull. L.M.S. **1** (1969) 79–88.

[Con69b] Conway, John: A characterization of Leech's lattice, Invent. Math. **7** (1969) 137–142.

[Con71] Conway, John: Three Lectures on Exceptional Groups in Higman-Powell, *Finite Simple Groups*. Academic Press, London 1971.

[Con.84] John H. Conway: A simplified construction for the Fischer-Griess monster group. Inventiones Mathematicae, 1984.

[CGP] Conway, J., M. Guy and N. Patterson: The Characters and Conjugacy Classes of .0, .1, .2, .3 and Suz. Unpublished (early 70s).

[CN] John H. Conway and Simon Norton: Monstrous Moonshine. Bull. London Math. Soc. **11** (1979) 308–339.

[Con-Sl], John H. Conway and Neal Sloane: Sphere Packings, Lattices and Groups. Grundlehren der mathematischen Wissenschaften 290. Springer Verlag, 1988.

[CM] Coxeter, H.S.M. and Moser: Generators and Relations for Discrete Groups. Springer, Berlin Heidelberg New York 1972.

[Curr] Curran, Peter: Fixed Point Free Action on a Class of Abelian Groups. Proc. Amer. Math. Soc. **57** (1976) 189–193.

[Cur76] Robert Curtis: A new combinatorial approach to M_{24}. Mathematics Proceedings of the Cambridge Philosophical Society 79 (1976) 25–42.

[Cur77] Robert Curtis: The maximal subgroups of M_{24}. Mathematics Proceedings of the Cambridge Philosophical Society 81 (1977), 185–192.

[CR] Curtis, Charles and Irving Reiner: *Representation Theory of Finite Groups and Associate Algebras*. Interscience, New York 1962.

[De] Ulrich Dempwolff: On extensions of an elementary abelian group of order 2^5 by $GL(5,2)$. Rend. Sem. Mat. Univ. Oadova **48** (1973) 359–364.

[Dic] Dickson, L.E.: *Linear Groups*. Dover, New York 1968.

[Die] Jean Dieudonné: *La Géométrie des Groups Classiques*. Springer, Berlin Heidelberg New York 1971.

[FT] Walter Feit and John Thompson: Solvability of groups of odd order. Pacific Journal of Mathematics **13** (1963) 775–1029.

[Fen] Daniel Fendel: A characterization of Conway's group ·3. J. Algebra **24** (1973) 159–196.

[Fi69] Fischer, Bernd: Finite Groups Generated by 3-Transpositions. Univ. of Warwick. Preprint 1969.

[Fi71a] Fischer, Bernd: Finite Groups generated by 3-Transpositions. Inventiones Math. **13** (1971) 232–246.

[Fi71b] Bernd Fischer: Private correspondence (a note on ${}^2E_6(2)$), 1971.

[Fi71c] Fischer, Bernd: Private correspondence, 1971.

[Fi73] Fischer, Bernd: Privately circulated note on F_2, 1973.

[FLT] Fischer, Bernd, D. Livingstone and Thorne: The Character Table of F_1. Unpublished.

[Froh] Daniel Frohardt: A trilinear form for the third Janko group. Jour. of Algebra **83** (1983) 349–379.

[FLM] I. Frenkel, J. Lepowsky and A. Meurman: Vertex Operator Algebras and the Monster. Academic Press, 1988.

[GaGa] Steve Gagola and Sidney Garrison: Real characters, double covers and the multiplier. J. of Algebra **74** (1982) 20–51.

[Go68] Daniel Gorenstein: Finite Groups. Harper and Row, New York 1968; 2nd ed. Chelsea, New York 1980.

[Go82] Daniel Gorenstein: Finite Simple Groups, An Introduction to their Classification. Plenum Press, New York 1982.

[GLS] Daniel Gorenstein, Richard Lyons and Ronald Solomon: The Classification of the Finite Simple Groups. Mathematical Surveys and Monographs **40**, number 1, The American Mathematical Society, 1994.

[GW] Gorenstein, D. and J. Walter: The Characterization of Finite Groups with Dihedral Sylow 2-Subgroups, I, II and III. Jour. of Alge. **2** (1965) 85–151; 218–270; 354–393.

[Gr72a] Robert L. Griess, Jr.: A sufficient condition for a finite group of even order to have nontrivial Schur multiplicator. Notices of the American Mathematical Society, 1972.

[Gr72b] Griess, Robert L., Jr.: Schur Multipliers of the Known Finite Simple Groups I. Bull. Amer. Math. Soc. **78** (1972) 78–71.

[Gr73a] Griess, Robert L., Jr.: Automorphisms of Extra Special Groups and Nonvanishing Degree 2 Cohomology. Pacific Jour. of Math. **48** (1973) 403–422.

[Gr73b] Griess, Robert L., Jr.: Schur Multipliers of Finite Simple Groups of Lie Type. Tran. Amer. Math. Soc. **183** (1973) 355–421.

[Gr74] Griess, Robert L., Jr.: Schur Multipliers of Some Sporadic Simple Groups. J. of Algebra **32** (1974) 445–466.

[Gr76a] Griess, Robert L., Jr.: The Structure of the "Monster" Simple Group. In: W. Scott and F. Gross (eds.), Proc. of the Conference on Finite Groups. Acad. Press, New York 1976, pp. 113–118.

[Gr76b] Griess, Robert L., Jr.: On a subgroup of order $2^{15}|GL(5,2)|$ in $E_8(\mathbf{C})$, the Dempwolff group and $Aut(D_8 \circ D_8 \circ D_8)$. J. Algebra **40** (1976) 271–279.

[Gr80a] The Schur multiplier of M_{22} and the associated component problem. Abstracts AMS **1** (1980), 80T-A40.

[Gr80b] Griess, Robert L., Jr.: Schur Multipliers of the Known Finite Simple Groups II. In the Santa Cruz Conference of Finite Groups, ed. by B. Cooperstein and G. Mason. Amer. Math. Soc., Providence 1980.

[Gr81] Griess, Robert L., Jr.: A Construction of F_1 as Automorphisms of a 198683 Dimensional Algebra. Proc. Nat. Acad. USA **78** (1981) 689–691.

[Gr82] Griess, Robert L., Jr.: The Friendly Giant. Invent. Math. **69** (1982) 1–102.

[Gr85] Robert L. Griess, Jr.: Schur multipliers of the known finite simple groups, III. In: Proceedings of the Rutgers Group Theory Year, 1983/84. University Press, Cambridge 1985, pp. 69–80

[Gr86a] Robert L. Griess, Jr.: Code Loops, Journal of Algebra **100** (1986) 224–234.

[Gr86b] Robert L. Griess, Jr.: The monster and its nonassociative algebra. Contemporary Mathematics **45** (1985) 121–157, American Mathematical Society.

[Gr87a] Griess, Robert L., Jr.: Sporadic groups, code loops and nonvanishing cohomology. Journal of Pure and Applied Algebra **44** (1987) 191–214.

[Gr87b] Griess, Robert L., Jr.: The Schur multiplier of McLaughlin's simple group. Arch. Math. **48** (1987) 31.

[Gr88] R. Griess: Code loops and a large finite subgroup containing triality for D_4; from Atti Conv. Inter. Teoria Geom. Combin. Firenze, Oct. 1986, Supplemento as Rend. Circ. Mat. Palermo **19** (1988) 79–98.

[Gr90] R. Griess: A Moufang loop, the exceptional Jordan algebra and a cubic form in 27 variables. J. Alg. **131** (1990) 281–293.

[Gr91] R. Griess: Elementary abelian p-subgroups of algebraic groups. Geometriae Dedicata **39** (1991) 253–305.

[Gr93] R. Griess: Codes, loops and p-locals. Article for proceedings of the Monster Bash conference, Columbus, Ohio, Ohio State University, May 20–23, 1993.

[GMS] R. Griess, U. Meierfrankenfeld and Y. Segev: A uniqueness proof for the Monster. Annals of Mathematics **130** (1989) 567–602.

[GrRy] R. Griess and A. J. E. Ryba: Embeddings of $U_3(8)$ and the Rudvalis group in algebraic groups of type E_7. Inventiones mathematicae **116** (1994) 215–242.

[Gru] Karl W. Gruenberg: Cohomological Topics in Group Theory. Lecture Notes in Mathematics, Vol. 143, Springer Verlag, 1970.

[HW] Hall, Marshall, Jr. and David Wales: The Simple Group of Order 604, 800. J. of Algebra **9** (1968) 417–450.

[Ha76] Harada, Koichiro: On the Simple Group F of Order $2^{14}3^65^6.7.11.19$. In: W. Scott and F. Gross (eds.), Proc. of a Conference on Finite Groups. Acad. Press, New York 1976, pp. 119–195.

[Ha78] The automorphism group and Schur multiplier of the simple group of order $2^{14}3^65^67.11.19$, Osaka Jour. of Math. **15** (1978) 3, 633–635.

[Held74] Dieter Held: A characterization of the alternating groups of degrees 8 and 9. J. Algebra **31** (1974) 91–116.

[Held69] Held, Dieter: The Simple Groups Related to M_{24}. J. Algebra **13** (1969) 253–296.

[HiSt] Peter Hilton and Urs Stammbach: A Course in Homological Algebra, Graduate Texts in Mathematics, Springer Verlag, Berlin, 1970.

[Hig62] Donald Higman: Flag-transitive collineation groups of finite projective spaces, Illinois J. Math. **6** (1962) 434–446.

[Hig64] Donald Higman: Finite Permutation Groups of Rank 3. Math. Z. **86** (1964) 145–156.

[Hig71] Higman, D.G.: A Survey of Some Questions and Results about Rank 3 Permutation Groups. Actes, Congres Intern, Math,. Tome 1, 361–365. Gauthier-Villars, Paris 1971.

[HM] G. Higman and J. McKay: On Janko's simple group of order 50,232,960. Bull. London Math. Soc. **1** (1969) 89–94, 219, and in "The Theory of Finite Groups", Brauer and Sah (eds.), pp. 65–77, Benjamin (1969).

[Ho] Chat-Yin Ho: A new 7-local subgroup of the monster. Journal of Algebra **115** (2) (1988) 513–520.

[Holt] Derek Holt: An interpretation of the cohomology groups $H^n(G, M)$. Journal of Algebra **60**, 2 (1979) 307.

[Hsu], Timothy Hsu: Class 2 and small Frattini Moufang loops (preprint, 1996); Explicit constructions of code loops as centrally twisted products (preprint, 1997).

[Hum] James Humphreys: Introduction to Lie Algebras and Representation Theory. Springer Verlag, Graduate Texts in Mathematics 9, 1972; third printing 1980.

[Jac] Jacobson, N.: *The Theory of Rings*. Amer. Math. Soc., Prov., 1943.

[Ja65] Zvonimir Janko: A new finite simple group with Abelian Sylow 2-subgroups. Proc. Nat. Acad. Sci. USA (1965) 657–658.

[Ja66a] Zvonimir Janko: A new finite simple group with Abelian Sylow 2-subgroups and its characterization. Jour. Algebra **3** (1966) 147–186.

[Ja66b] Janko, Z. and J. G. Thompson: On a Class of Finite Simple Groups of Ree, J. of Algebra **4** (1966) 274–292.

[Ja67] Zvonimir Janko: A characterization of a new simple group, in "The Theory of Groups", Kovacs and Neumann (eds.), pp. 205–208, Gordon and Breach, London and New York 1967.

[Ja68] Janko, Z.: Some New Simple Groups of Finite Order I. First Naz. Alta Math. Symposia Math **1** (1968) 25–65.

[Ja76] Janko, Z.: A new finite simple group of order 86,775,570,046,077,562,880 which posesses M_{24} and the full covering group of M_{22} as subgroups. Journal of Algebra **42** (1976) 564–596.

[JW69] Zvonimir Janko and S.K. Wong: A characterization of the Higman-Sims simple group. J. Algebra **13** (1969) 517–534.

[JW71] Zvonimir Janko and S. K. Wong: A characterization of McLaughlin's simple group. J. Algebra **20** (1971) 203–255.

[Jo] Peter Johnson: Covering loops of abelian groups. To appear in Journal of Algebra.

[Le64] Leech, John: Some Sphere Packings in Higher Space. Can. J. Math **16** (1964) 657–682.

[Le67] Leech, John: Notes on Sphere Packings. Can. J. Math. **19** (1967) 251–267.

[LS] J. Leon and C. Sims: The existence and uniqueness of a simple group generated by $\{3,4\}$-transpositions. Bull. Amer. Math. Soc. **83** (1977) 1039–1040.

[Lin69] John Lindsay: Linear groups of degree 6 and the Hall-Janko group, article in Theory of Finite Groups, R. Brauer and C.H. Sah (eds.). Benjamin, New York 1969.

[Lin71a] John Lindsay: A correlation between $PSU(4,3)$, the Suzuki group and the Conway group. Transactions of the American Mathematical Society **157** (1971) 189–204.

[Lin71b] John Lindsay: On the Suzuki and Conway groups. Mathematics Proceedings of the Cambridge Philosophical Society **21** (1971) 107–109.

[Ly] Richard Lyons: Evidence for a new finite simple group. Jour. of Algebra **20** (1972) 540–569; and **34** (1975) 188–189.

[MacL] Saunders MacLane: Homology. Grundlehren series, Band 114, Springer Verlag, 1967.

[MacW-Sl] MacWilliams, J. and Sloane, N.: The Theory of Error-Correcting Codes. North Holland, Amsterdam, second reprint 1983.

[Ma84] Geoffrey Mason: Frame shapes and rational characters of finite groups. Journal of Algebra, 1984.

[Ma85] Geoffrey Mason: M_{24} and certain automorphic forms. Contemporary Mathematics **45** (1985) 223–244; American Mathematical Society, Providence.

[Mat] Mathieu, Emil: Memoire sur la nombre de valeurs que peut acquirer une fonction quand on y permut ses variables de toutes les manieres possibles, Liouville's J. **5** (1860) 9–42; Memoire sur l'etude des fonctions de plusieurs quantities, sur la maniere de les formas et sur les substitutions qui les laissent invariables, Liouville's J. **6** (1861) 241–323; Sur la fonction cinq fois transitives des 24 quantites, Liouville's J. **18** (1873) 25–46.

[Ma79] Pierre Mazet: Sur le multiplicator de Schur du groupes de Mathieu, C.R. Acad. Sci. Paris, Serie A-B, **289**, 14 (1979) A659–661.

[Ma82] Pierre Mazet: Sur le multiplicator de Schur du groupes de Mathieu. Journal of Algebra **77**, 2 (1982) 552–576.

[McK] J. McKay: Computing with finite simple groups. In: Proceedings of the Second International Conference on the Theory of Groups, pp. 448–452.

[McL69] McLaughlin, Jack: A simple group of order 898,128,000, article in Finite Simple Groups, ed. by R. Brauer and C. H. Sah, pp. 109–112, Benjamin, 1969.

[McL75] Jack McLaughlin: Course notes for cohomology of groups. University of Michigan, 1975.

[HM] Milnor, J. and Husemoller, D.: Symmetric Bilinear Forms. Springer, Berlin Heidelberg New York 1973.

[Nie67] Niemeyer, H. V.: Definite Quadratische Formen der Diskriminante 1 und Dimension 24. Doctoral Dissertation, Göttingen 1968.

[No75] Simon Norton: F and other simple groups. Ph.D. Thesis, University of Cambridge, 1975.

[No80] Simon Norton: The construction of J_4. In: The Santa Cruz Conference on Finite Groups, Cooperstein and Mason (eds.). American Mathematical Society, 1980, pp. 171–178.

[Ogg] Andrew Ogg: Modular Forms and Dirichlet Series, W. A. Benjamin, Inc., 1969.

[O'N] Michael O'Nan: Some evidence for the existence of a new simple group, Proc. London Math. Soc. **32** (1976) 421–479.

[Pat] Patterson, Nicholas J.: On Conway's Group .0 and Some Subgroups. Thesis, University of Cambridge, 1974.

[PW76a] Nickolas J. Patterson and S. K. Wong: The nonexistence of a certain simple group. Journal of Algebra **39** (1976), 138–149.

[PW76b] Nickolas J. Patterson and S. K. Wong: A characterization of the Suzuki sporadic simple group. Journal of Algebra **39** (1976) 277–286.

[Po] Harriet K. Pollatsek: Cohomology Groups of Some Linear Groups over Fields of Characteristics 2, Ill. J. of Math. **15** (1971) 393–417.

[Rich91] Tom Richardson: Local subgroups of the monster and odd code loops. PhD. Thesis, University of Michigan, 1991.

[Rich92] Tom Richardson: Elementary abelian subgroups of the monster, Group Theory (Granville, HO), 244–249, World Scientific Publications, River Edge, N.J., 1992.

[Rich94] Tom Richardson: Elementary abelian 5-subgroups of the monster. Journal of Algebra **165** (1994), 223–241.

[Rich95] Tom Richardson: Local subgroups of the monster and odd code loops. Transactions of the American Mathematical Society **347** (1995), 1453–1531.

[Rob] Derek Robinson: The Vanishing of Certain Homology and Cohomology Groups. J. of Pure and Appl. Algebra **7** (1976) 145–167.

[RoSm] Ronan, Mark and Steven D. Smith: 2-Local Geometries for Finite Groups. Article in the Santa Cruz Conference on Finite Groups. Amer. Math. Soc., Prov. 1980.

[Rot] Joseph Rotman: Notes on Homological Algebra. Van Nostrand 1970.

[Ru] Arunas Rudvalis: A rank 3 simple group of order $2^{14}3^3 5^3 7.13.29$, I, Journal of Algebra **86** (1984) 1, 181–218; A rank 3 simple group of order $2^{14}3^3 5^3 7.13.29$, II: characters of G and $\hat{G}$, Journal of Algebra **86**, 1 (1984) 219–258.

[Schur04] Schur, I.: Über die Darstellung der endlichen Gruppen durch gebrochene lineare Substitutionen. J. Reine und Angew. Math. **127** (1904) 20–50.

[Schur07] Schur, I: Über die Darstellung der endlichen Gruppen durch gebrochene lineare Substitutionen. J. Reine und Angew. Math. **132** (1907) 85–107.

[Schur11] Schur, I: Über die Darstellung der symmetrischen und alternienden Gruppen durch gebrochene lineare Substitutionen. J. Reine und Angew. Math. **139** (1911) 155–250.

[Serre] Jean-Pierre Serre: A Course in Arithmetic. Springer Verlag, Graduate Texts in Mathematics 7, 1973.

[Shi] Goro Shimura: Introduction to the Arithmetic Theory of Automorphic Functions. Iwanami Shoten Publishers and Princeton University Press, 1971.

[Sims72] Charles C. Sims: The existence and uniqueness of Lyons' group. In: Finite Groups '72, Gagen, Hales and Shult (eds.). North Holland, Amsterdam, 1973, pp. 138–141.

[Sims0] Charles C. Sims: A construction of O'Nan's group. Not published.

[SmiF] Fred Smith: A characterization of Conway's simple group ·2, J. Algebra **31** (1974), 91–116.

[Smi] Smith, Steve: Nonassociative Commutative Algebras for Triple Covers of 3-Transposition Groups. Michigan Math. J. **24** (1977) 273–287.

[Sol] Ronald Solomon: On the finite simple groups and their classifications. Notices of the American Mathematical Society **42**, 2 (1995) 231–239.

[St59] Steinberg, R.: Variations on a Theme of Chevalley. Pacific J. of Math. **9** (1959) 875–891.

[St62] Robert Steinberg: Générateurs, relations, revêtements de groupes algébriques. Colloq. Théorie des Groupes Algébriques (Bruxelles, 1962), Librairie Universitaire, Louvain (Gauthier-Villars, Paris, 1962) pp. 113–127.

[St63] Steinberg, R.: Representation of Algebraic Groups. Nagoya Math. J. **22** (1963) 33–56.

[St67] Steinberg, R.: Lectures on Chevalley Groups. Lecture Notes, Math. Dept., Yale Univ., 1967.

[St81] Robert Steinberg: Generators, Relations and Coverings of Algebraic Groups, II. Journal of Algebra **71** (1981), 527–543.

[Stell] Bernd Stellmacher: Enifache Gruppen, die von einer Konjugiertenklasses von Elementen der Ordnung drei erzeugt werden. Journal of Algebra **30** (1974) 320-354.

[Suz62] Suzuki, M.: On a Class of Doubly Transitive Groups; I, II. Annals of Math. **75** (1962) 105–145; **79** (1964) 514–589.

[Suz68] Michio Suzuki: A simple group of order 448,345,497,600. Article in R. Brauer, C.-H. Sah (eds.): Theory of Finite Groups. W.A. Benjamin, Inc. 1969, pp. 113–120.

[Suz72] Michio Suzuki: Lecture at University of Florida, Gainesville, March, 1972.

[Thomas] Gomer Thomas: A characterization of the groups $G_2(2^n)$, Journal of Algebra **13** (1969), 87–118.

[Th76] J. Thompson: A simple subgroup of $E_8(3)$. In: Finite Groups, Iwahori, N. (ed.). Japan Society for Promotion of Science, Tokyo, 1976, pp. 113–116.

[Th79a] Thompson, J.G.: The Uniqueness of the Fischer-Griess Monster. Bull. London Math. Soc. **11** (1979) 340–346.

[Th79b] J. Thompson: Some numerology between the Fischer-Griess Monster and the elliptic modular function. Bull. London Math. Soc. **11** (1979), 352–353.

[Timm75] Timmesfeld, Franz: Groups Generated by Root Involutions, I, II. J. of Algebra **33** (1975) 75–135; **35** (1975) 367–441.

[Timm80] Timmesfeld, Franz: Groups of $GF(2)$-type and Realted Problems. Article in *Finite Simple Groups II*, ed. by Michael J. Collins. Acad. Press, London 1980.

[Tits64] Jacques Tits: Algebraic and abstract simple groups. Annals of Math **80** (1964).

[Tits80a] Jacques Tits: Four Presentations of Leech's lattice. In: Finite Simple Groups, II. Proceedings of a London Math. Soc. Research Symposium, Durham, 1978, ed. by M. J. Colllins. Adademic Press, London, New York, 1980, pp. 306–307

[Tits80b] Jacques Tits: Quaternions over $\mathbf{Q}(\sqrt{5})$, Leech's lattice and the sporadic group of Hall-Janko. J. Algebra **63** (1980), 56–75.

[Tits83a] Jacques Tits: Résumé de cours, Annuaire de Collège de France, 1982-83.

[Tits83b] Jacques Tits: Remarks on Griess' construction of the Griess-Fischer sporadic group, I, II, III, IV, lettres polycopiée, 1983.

[Tits83c] Jacques Tits: Le Monstre [d'après R. Griess, B. Fischer, et al.]. Séminaire Bourbaki, Novembre 1983, No. 620.

[Tits84] Jacques Tits: On R. Griess' "Friendly giant". Inventiones mathematicae **78**, (1984) 491–499.

[Todd] John Todd: A Representation of the Mathieu Group M_{24} as a Collineation Group. Annali di Math. Pura ed. Applicata **71** (1966) 199–238.

[Ven78] B.B. Venkov: The classification of integral even unimodular 24-dimensional quadratic forms. Trudy Maatematicheskogo Instituta imeni V. A. Steklova **148** (1978), 65–76 = Proceedings of the Steklov Institute of Mathematics (No. 4, 1980)

[Wales] David Wales: Uniqueness of the graph of a rank three group. Pacific Journal of Mathematics, Vol. 30, No. 1, 1969.

[Wie] Helmut Wielandt: Finite Permutation Groups. Academic Press, New York London 1964.

[Wil86] R. Wilson: Is J_1 a subgroup of the monster? Bulletin London Math. Soc. **18**, 4 (1986) 349–350.

[Wil88] Robert A. Wilson: On the 3-local subgroups of Conway's group Co_1. Journal of Algebra **113** (1988) 261–262.

[Witt37] Ernst Witt: Theorie der quadratischedn Formen in beliebigen Körpern. JRAM **176** (1937), 31–44.

[Witt38a] Ernst Witt: Die 5-fach transitiven Gruppen von Mathieu. Hamburg Universität Abhandlungen aus dem Mathematischen Seminar **12** (1938) 256–264.

[Witt38b] Ernst Witt: Über Steinersche Systeme. Hamburg Universität Abhandlungen aus dem Mathematischen Seminar **12** (1938) 265–275.

Index

We list at least one significant occurrence for an entry.

Numerical or special Symbols

List of Group Theoretic Notations

This book uses notations and definitions which have been in general usage among finite group theorists since the sixties decade. See [Go68] and [Hu]. For convenience, we list a few group-theoretic notations below.

Notation	Meaning	In this text
x^y	$y^{-1}xy$, for group elements x, y	
$[x, y]$	the commutator $x^{-1}y^{-1}xy$ of group elements x, y	
$G_{x_1, x_2, \ldots}$	the stabilizer in the group G of points x_1 and x_2 and ...	(1.1), (1.3)
$N_G(P)$, $C_G(P)$	P is a subset of the group G, $C_G(P) = \{x \in G \mid xy = yx, \text{ for all } y \in P\}$ and $N_G(P) = \{x \in G \mid x^{-1}Px = P\}$	(1.5) and throughout
$Stab_G(A)$	G acts on the set X and $A \subseteq X$; this is the subgroup of elements of G which take A onto A (see above)	Chapter 6
$O_p(G)$	the largest normal p-subgroup of the group G (p is a prime)	(2.15)
$O_{p'}(G)$	the largest normal p'-subgroup of the group G	(1.11)
W_{E_8}, W_Φ	the Weyl group of type E_8, type Φ (some root system)	
$A.B, A{:}B, A \cdot B$	a group extension, a split extension, a nonsplit extension	(2.1)
Sym_n, Σ_n	the symmetric group of degree n	
Alt_n	the alternating group of degree n	
$Perm(n, F)$, $Mon(n, F)$, $Diag(n, F)$, etc.	various groups of monomial matrices	(3.12)

The Orders of the Finite Simple Groups

Alternating groups		
$A_n,\ n \geq 5$	$\frac{1}{2}(n!)$	
Group G of Lie type	**Order of G**	**d**
$A_n(q)$	$q^{n(n+1)/2} \prod_{i=1}^{n} (q^{i+1} - 1)$	$(n+1, q-1)$
$B_n(q),\ n \geq 1$	$q^{n^2} \prod_{i=1}^{n} (q^{2i} - 1)$	$(2, q-1)$
$C_n(q),\ n > 2$	$q^{n^2} \prod_{i=1}^{n} (q^{2i} - 1)$	$(2, q-1)$
$D_n(q),\ n > 3$	$q^{n(n-1)}(q^n - 1) \prod_{i=1}^{n-1} (q^{2i} - 1)$	$(4, q^n - 1)$
$G_n(q)$	$q^6(q^6 - 1)(q^2 - 1)$	1
$F_4(q)$	$q^{24}(q^{12} - 1)(q^8 - 1)(q^6 - 1)(q^2 - 1)$	1
$E_6(q)$	$q^{36}(q^{12} - 1)(q^9 - 1)(q^8 - 1)(q^6 - 1)(q^5 - 1)(q^2 - 1)$	$(3, q-1)$
$E_7(q)$	$q^{63}(q^{18} - 1)(q^{14} - 1)(q^{12} - 1)(q^{10} - 1)(q^8 - 1)(q^6 - 1)(q^2 - 1)$	$(2, q-1)$
$E_8(q)$	$q^{120}(q^{30} - 1)(q^{24} - 1)(q^{20} - 1)(q^{18} - 1)(q^{14} - 1)(q^{12} - 1)(q^8 - 1)(q^2 - 1)$	1
${}^2A_n(q),\ n > 1$	$q^{n(n+1)/2} \prod_{i=1}^{n} (q^{i+1} - (-1)^{i+1})$	$(n+1, q+1)$
${}^2B_2(q),\ q = 2^{2m+1}$	$q^2(q^2 + 1)(q - 1)$	1
${}^2D_n(q),\ n > 3$	$q^{n(n-1)}(q^n + 1) \prod_{i=1}^{n-1} (q^{2i} - 1)$	$(4, q^n + 1)$
${}^3D_4(q)$	$q^{12}(q^8 + q^4 + 1)(q^6 - 1)(q^2 - 1)$	1
${}^2G_2(q),\ q = 3^{2m+1}$	$q^3(q^3 + 1)(q - 1)$	1
${}^2F_4(q),\ q = 2^{2m+1}$	$q^{12}(q^6 + 1)(q^4 - 1)(q^3 + 1)(q - 1)$	1
${}^2E_6(q)$	$q^{36}(q^{12} - 1)(q^9 + 1)(q^8 - 1)(q^6 - 1)(q^5 + 1)(q^2 - 1)$	$(3, q+1)$

Sporadic Groups

Group	Order	Group	Order
M_{11}	$7920 = 2^4 \cdot 3^2 \cdot 5 \cdot 11$	He	$2^{10}3^3 5^2 \cdot 7^3 \cdot 17$
M_{12}	$95040 = 2^6 \cdot 3^3 \cdot 5 \cdot 11$	Ly	$2^8 3^7 5^6 7 \cdot 11 \cdot 31 \cdot 37 \cdot 67$
M_{22}	$443520 = 2^7 \cdot 3^2 \cdot 5 \cdot 7 \cdot 11$	ON	$2^9 3^4 5 \cdot 7^3 \cdot 11 \cdot 19 \cdot 31$
M_{23}	$10200960 = 2^7 \cdot 3^2 \cdot 5 \cdot 7 \cdot 11 \cdot 23$	Co_1	$2^{21}3^9 5^4 7^2 11 \cdot 13 \cdot 23$
M_{24}	$244823040 = 2^{10} \cdot 3^3 \cdot 5 \cdot 7 \cdot 11 \cdot 23$	Co_2	$2^{18}3^6 5^3 7 \cdot 11 \cdot 23$
J_1	$175560 = 2^3 \cdot 3 \cdot 5 \cdot 7 \cdot 11 \cdot 19$	Co_3	$2^{10}3^7 5^3 7 \cdot 11 \cdot 23$
$HJ = J_2$	$2^7 \cdot 3^3 \cdot 5^2 \cdot 7$	Fi_{22}	$2^{17}3^9 5^2 7 \cdot 11 \cdot 23$
J_3	$2^7 \cdot 3^5 \cdot 5 \cdot 17 \cdot 19$	Fi_{23}	$2^{18}3^{13}5^2 7 \cdot 11 \cdot 13 \cdot 17 \cdot 23$
J_4	$2^{21} \cdot 3^3 \cdot 5 \cdot 7 \cdot 11^3 \cdot 23 \cdot 29 \cdot 31 \cdot 37 \cdot 43$	Fi_{24}	$2^{21}3^{16}5^2 7^3 \cdot 11 \cdot 13 \cdot 17 \cdot 23 \cdot 29$
HS	$2^9 3^2 5^3 \cdot 7 \cdot 11$	F_5	$2^{15}3^{10}5^3 7^2 13 \cdot 19 \cdot 31$
McL	$2^7 3^6 5^3 \cdot 7 \cdot 11$	F_3	$2^{14}3^6 5^6 \cdot 7 \cdot 11 \cdot 19$
Suz	$2^{13}3^7 5^2 7 \cdot 11 \cdot 23$	F_2	$2^{41}3^{13}5^6 7^2 11 \cdot 13 \cdot 17 \cdot 19 \cdot 23 \cdot 31 \cdot 47$
Ru	$2^{14}3^3 5^3 7 \cdot 13 \cdot 29$	$F_1 = IM$	$2^{46}3^{20}5^9 \cdot 7^6 \cdot 11^2 \cdot 13^3 \cdot 17 \cdot 19 \cdot 23 \cdot 29 \cdot 31 \cdot 41 \cdot 47 \cdot 59 \cdot 71$